GROUNDING AND SHIELDING TECHNIQUES

GROUNDING AND SHIELDING TECHNIQUES

Fourth Edition

RALPH MORRISON

A Wiley-Interscience Publication
John Wiley & Sons, Inc.
New York • Chichester • Weinheim • Brisbane • Singapore • Toronto

This book is printed on acid-free paper. ⊗

Copyright © 1998 by John Wiley & Sons, Inc. All rights reserved.

Published simultaneously in Canada.

No part of this publication may be reproduced, stored in a retrieval system or transmitted in any form or by any means, electronic, mechanical, photocopying, recording, scanning or otherwise, except as permitted under Section 107 or 108 of the 1976 United States Copyright Act, without either the prior written permission of the Publisher, or authorization through payment of the appropriate per-copy fee to the Copyright Clearance Center, 222 Rosewood Drive, Danvers, MA 01923, (508) 750-8400, fax (508) 750-4744. Requests to the Publisher for permission should be addressed to the Permissions Department, John Wiley & Sons, Inc., 605 Third Avenue, New York, NY 10158-0012, (212) 850-6011, fax (212) 850-6008, E-Mail: PERMREQ @ WILEY.COM.

Library of Congress Cataloging-in-Publication Data:

Morrison, Ralph.
    Grounding and shielding techniques / Ralph Morrison.—
4th ed.
      p.   cm.
    Includes index.
    ISBN 0-471-24518-6 (cloth : alk. paper)
    1. Electronic instruments—Protection.   2. Electric currents—
Grounding.   3. Shielding (Electricity)   I. Title.
TK7878.4.M66   1998
621.3815′4—dc21
                                                                                    97-26394
                                                                                       CIP

Printed in the United States of America.

10 9 8 7 6 5 4 3 2 1

CONTENTS

Preface xi

1 Electrostatics 1

    1.1   Introduction, 1
    1.2   The Low-Frequency Signal Definition, 3
    1.3   Charge, 4
    1.4   Forces Between Charges, 4
    1.5   Electric Field, 5
    1.6   The E Field for Charge Distributions, 6
    1.7   The Concept of Voltage, 6
    1.8   Voltage Gradient, 7
    1.9   Spherical Conductor with a Charge, 8
    1.10 Capacitance, 9
    1.11 The Displacement Field D, 9
    1.12 Field Representations, 10
    1.13 Points of Difficulty, 12
    1.14 MKS System of Units, 12
    1.15 Charges on Spherical Shells, 14
    1.16 The Earth Plane, 15
    1.17 Typical Charge Distributions, 16
    1.18 Cylindrical Surfaces, 17
    1.19 Parallel Plate Capacitors, 19
    1.20 Electric Field Energy, 20
    1.21 Self- and Mutual Capacitance, 21
    1.22 Example of Mutual Capacitance, 22

1.23 Importance of Electric Field Concepts, 22
1.24 A Powerful Tool, 23

2 Magnetics 25

2.1 Introduction, 25
2.2 Lines of Force and Flux, 26
2.3 The B Field, 26
2.4 The H Field, 27
2.5 Faraday's Law, 29
2.6 The Magnetic Circuit, 29
2.7 The Transformer, 31
2.8 Hysteresis—Magnetic Materials, 33
2.9 Inductance, 35
2.10 The Inductance of a Solenoid, 35
2.11 Energy in the Magnetic Field, 36
2.12 Magnetic Units, 36
2.13 Leakage Inductance, 37
2.14 The Audio Transformer, 39
2.15 Magnetic Storms, 39
2.16 Spacecraft Potentials, 40
2.17 The E and H Fields Together, 40
2.18 Fields and Components, 41
2.19 The Inductance of Isolated Conductors, 42

3 The Transport of Signals and Power 43

3.1 Introduction, 43
3.2 The Transmission Line, 44
3.3 Transmission Line Terminations, 45
3.4 Transmission Line Fields, 47
3.5 Ground Planes, 49
3.6 Coaxial Transmission Lines, 50
3.7 Sine Waves and Transmission Lines, 50
3.8 Terminations, 53
3.9 Poynting's Vector, 54
3.10 Radiation, 55
3.11 Radiation from a Dipole, 56
3.12 Radiation from a Current Loop, 57
3.13 Effective Radiating Power, 59

Contents vii

4 Fields and Conductors — 61

4.1 Introduction, 61
4.2 Ohms per Square, 62
4.3 Reflection, 64
4.4 Skin Effect, 64
4.5 Shielding Effectivity, 65
4.6 Apertures, 66
4.7 Independent and Dependent Apertures, 67
4.8 Closing Apertures, 67
4.9 Waveguides Beyond Cutoff, 68
4.10 A Review of Fields Entering an Enclosure, 69
4.11 The Coupling of Fields to Circuits, 69
4.12 The Fields in a Room, 71
4.13 Common Mode and Normal Mode, 73

5 Electrostatic Shielding—I — 75

5.1 Introduction, 75
5.2 The Circuit in a "Box", 76
5.3 External Grounding of One Conductor, 78
5.4 The Game, 78
5.5 The Transformer Connection, 81
5.6 The Single Transformer Shield, 82
5.7 The Three-Shield Solution, 84
5.8 The Single-Ended Instrument, 85
5.9 Single-Ended Signal Sources, 86
5.10 Balanced Signal Sources, 87
5.11 The Two-Ground Problem, 88
5.12 The Fundamental Instrumentation Problem, 90
5.13 The Differential Amplifier, 91

6 Electrostatic Shielding—II — 93

6.1 Shielding Low-Frequency Signals, 93
6.2 Shielding Single-Ended Signals, 94
6.3 Shielding Balanced Sources, 96
6.4 Floating Circuits, 97
6.5 Shielding Grounded Unbalanced Circuits, 98
6.6 Shielding Electronic Equipment, 99
6.7 Shielding Instrumentation Amplifiers, 100

- 6.8 The Driven Shield, 103
- 6.9 The Analog Input Cable, 104
- 6.10 Aluminum Foil Shielding, 105
- 6.11 The Drain Wire, 105
- 6.12 Low-Noise Cable, 106
- 6.13 Reactive Coupling in Circuits, 107
- 6.14 Guard Rings, 108

7 Common-Impedance Coupling — 111

- 7.1 Introduction, 111
- 7.2 Amplifier Common Impedance, 112
- 7.3 Strain-Gage Excitation Supplies, 112
- 7.4 Transformer Shielding for Excitation Supplies, 116
- 7.5 Star Connections, 117
- 7.6 Neutrals as Common Impedances, 119
- 7.7 The Earth as a Common Impedance, 120
- 7.8 The Forward Referencing Amplifier, 122
- 7.9 Driven Power Stages (Isolated Output), 123

8 Circuit Designs — 127

- 8.1 Introduction, 127
- 8.2 Common-Mode Rejection—Postmodulation, 128
- 8.3 The Common-Mode Attenuator, 128
- 8.4 High-Input Impedance Circuits, 130
- 8.5 Balanced Excitation Supplies, 133
- 8.6 Charge Amplifiers, 134
- 8.7 Charge Amplifiers and Calibration, 137
- 8.8 Calibration Guarding, 137
- 8.9 Filtering Analog Signals, 138
- 8.10 The ac Amplifier, 140
- 8.11 Suppression Circuits, 140
- 8.12 Switching Regulators, 142
- 8.13 Off-Line Switchers, 144
- 8.14 Mounting Switching Transistors or FETs, 144
- 8.15 Parallel Active Components, 145
- 8.16 The Medical Problem, 145
- 8.17 The Mad Cow Problem, 148

Contents ix

9 Utility Power 149

9.1 Introduction, 149
9.2 The Ground Plane, 150
9.3 Single-Point Grounding of Ground Planes, 151
9.4 Racks as a Ground Plane, 152
9.5 Extensions of a Ground Plane, 153
9.6 The Earth as a Ground Plane, 153
9.7 Isolated Grounds, 154
9.8 Separately Derived Power, 155
9.9 Power Isolation Transformers, 156
9.10 Computer Power Centers, 157
9.11 Power-Line Filters, 158
9.12 Facility Power Filters, 159
9.13 Transient Power Loads, 160
9.14 Controllers, 161
9.15 Transient Protection, 162
9.16 Ungrounded Power, 163
9.17 The NEC and Power Connections, 163
9.18 Bending Magnetic Fields, 164

10 High-Frequency Design 167

10.1 Introduction, 167
10.2 The PC Board, 168
10.3 PC Ground Planes, 169
10.4 PC Board Transmission Lines, 169
10.5 PC Boards and Radiation, 170
10.6 PC Boards Decoupling Capacitors, 170
10.7 PC Boards—Ground Planes and Power Planes, 171
10.8 PC Boards—Ribbon Cable, 172
10.9 High-Frequency Transport and Open Cable, 173
10.10 High-Frequency Transport Over Coax, 174
10.11 Cable Transfer Impedance, 174
10.12 Cable Shield Terminations, 175
10.13 The Chattering Relay Test, 177
10.14 The Bond, 178
10.15 The Transport of High-Frequency Power, 179
10.16 Mixing Analog and Digital Signals, 179
10.17 The Sniffer, 179
10.18 Screen Rooms, 180
10.19 Screen Room Power Filters, 181

10.20 Grounding the Screen Room, 181
10.21 The Screen Room Design, 182
10.22 Using a Screen Room, 182

11 Pulses and Step Functions   185

11.1 Introduction, 185
11.2 Spectrum of a Square Wave, 186
11.3 Spectrum of a Single Event, 188
11.4 A Valuable Calculating Tool, 191
11.5 The ESD (Electrostatic Discharge), 191
11.6 ESD Protection, 192
11.7 ESD and Its Characteristics, 192
11.8 ESD Testing, 193

Index   195

PREFACE TO THE FOURTH EDITION

It has been 11 years since the third edition was published. In this period, electronics has seen an amazing growth. The growth direction has been to add digital processing, high-speed communication, and mass data storage. This change has brought the analog world into direct contact with high-speed operations. Analog designs are buried in the middle of hardware that must meet radiation and susceptibility standards. Often the close proximity creates new problems for the designer.

The third edition recognized the need to say something about the higher-frequency issues. This minimum treatment is now inadequate, and this has led me to undertake a rewrite of the entire book. The first edition stressed electrostatics that worked back in 1967. In 1997 it is an electromagnetic world, and the fourth edition needed to face this fact.

It is true that many issues can be resolved by looking at electrostatic processes only. This fact has not been overlooked. Chapters 5 and 6 treat issues that do not require an understanding of radiation or susceptibility. Note that these chapters follow a full treatment of fields, how they are generated, and how they enter equipment. The physics that is used is reduced to its simplest form to allow the reader to understand how and why things behave as they do. With a good foundation the reader should be able to see how a ground plane functions whether it is for a PC board or a computer facility.

The first edition stressed the idea that circuit theory is a poor match to many interference issues. The size of components, their position in a circuit, and the sequence of interconnection are all important. This idea is still true today except that more physics must be introduced. The added physics is kept simple, and the connection to real-world problems is maintained.

Power enters every piece of hardware. The fourth edition expands on issues associated with facility transformers as well as hardware transformers. The issues of filtering both a facility and hardware are discussed. The discussion of multishielded transformers has been expanded. Within the instrument the discussion of multishielding has been trimmed so that it is easier to follow. Today the emphasis is on how to get rid of expensive transformers. Electronics has progressed to the point where it is practical to limit the use of transformers to reduce both price and weight. The new issues that are raised are discussed in detail.

It has been a source of satisfaction to see *Grounding and Shielding Techniques* survive all of these years. Thirty years is indeed a long time, especially in electronics. My early practical experience allowed me write the first book. The book and all the resulting consulting and teaching opportunities have expanded my understanding of this topic. The fourth edition reflects much of this experience. Teaching a subject helps put a lot of loose ends to rest.

I do not fear the next question asked by an engineer, because I feel I understand the basics after all of these years. Most explanations are simple once the problem has been properly identified. When there are multiple problems, I must do a lot of probing. Most of my probing is related to an understanding of what are the right issues. Many engineers have trouble describing their problem. I must find out what is relevant and what is not, and I must find a way to draw the correct geometry. Experience eventually leads an engineer to separate all of the variables, but this takes time. Problems often crop up at the end of a project, and this is the worst possible time for change or experimentation. This is the time they often attend a class. When I give an answer that undermines the entire concept, there is one unhappy engineer. The next challenge is to find a way out of the dilemma. Often this is when the consultant is called in.

This book must be practiced. It takes faith to accept the idea that all electrical phenomena were explained in the last century (1800s). Looking for new physics or new rules is not the correct approach. Letting go of misconceptions is not easy. Our personal explanation is often our only shield against reality. We feel very vulnerable when our understanding is challenged. This book provides some of these challenges. It is up to the reader to read the facts, find the truth, and use the result to attack the next problem. I can only hope that I have been of help.

RALPH MORRISON

Eureka, California
December 1997

GROUNDING AND
SHIELDING TECHNIQUES

1

ELECTROSTATICS

1.1 INTRODUCTION

The control of electrical interference is an ongoing problem for engineers. The words *grounding* and *shielding* seem to suggest solutions to the difficulties most designers encounter. Unfortunately, words alone do not solve problems. It is hoped that this book will provide the necessary insight so that solutions can be found. The engineer usually turns to his or her own past experience and to the experience of others to propose solutions when there are difficulties. Often solutions fail, and this results in a great deal of trial and error. The engineer often has a hard time relating cause and effect, and this can result in some strange logic to explain the nature of the problem.

An analog circuit designer knows from experience that the circuit is less apt to pick up noise if the metal box surrounding the circuit is connected to a power supply common or circuit common. The designer also knows that connecting this same ground to a water pipe or to the power neutral can sometimes reduce noise pickup. The full nature of the coupling or why the noise is reduced is not always clear. The conclusion that is often reached is that a better ground or earth would solve many problems. This, in turn, leads to very fancy grounding rods and grounding grids. These rods or grids are connected to earth. A true understanding of what is happening does not come from this simple experiment.

The question that is often asked is, Why must we be earthed at all? The answer lies in one word—safety. The problem starts with the utility power supplied to a facility. Earthing the power system is necessary to provide lightning protection. In three-phase systems the neutral is earthed, and in single-phase systems one of the power conductors is earthed. This earthing takes place at a service entrance or at the secondary of a distribution transformer. The rules for this earthing are elaborate and are a part of the National Electrical Code (NEC). If there were no earth connection, lightning could follow power conductors into a facility and jump to the plumbing to get to earth. This poses both a fire hazard and a safety hazard.

The safety issue extends to all the metal in a facility. If a "hot" power conductor were to contact any building metal, there is the obvious chance of electrical shock. If this type of short were to occur, the power circuit must be opened. This means that all metal in a facility must be bonded together. Earth connections alone are inadequate because it is difficult to obtain resistances below 5 Ω. The resistance through two earth connections would be 10 Ω. On a 120-V circuit the fault current would be 12 A, and this is certainly within the circuit rating. Two things are now apparent. The power system must be earthed, and all metals* in a facility must be bonded together and also connected to the power system. The metal in any one facility is known as the *grounding electrode system.*

Finding a good ground or best ground or avoiding a ground used by someone else is a strange occupation. Even the words that are used are not a proper part of engineering. Terms such as *clean ground, quiet ground,* or *secure ground* are the result of a certain amount of frustration. These terms tend to mislead rather than help the engineer. They imply a quality which cannot be guaranteed. These terms suggest a circuit approach to solving problems. In most cases the problem can only be resolved by considering circuit geometry.

In the hope of avoiding difficulty, very elaborate grounding schemes are often designed into facilities. It is difficult to experiment with an entire facility, and thus facilities remain rather constant during their life. The biggest issues arise when a facility is to be expanded. Should the old thinking prevail, or should there be a new approach? This is a serious issue because expansions involve new and higher-speed devices and change is expensive. This higher speed implies greater bandwidth and a greater chance of radiation and associated interference. There is a suspicion that the old approach may not be optimum for this newer equipment.

The education of most electrical and electronic engineers revolves around circuit theory. Currents flow in conductors, there are voltage and

---

* These are metals that support the power wiring or might come in contact with power wiring.

current sources, and there is energy storage in capacitors and inductors. A very beautiful mathematical approach provides solutions to any given circuit problem. In advanced courses there is a mathematics to treat linear circuits when they are excited by nonsinusoid. The subjects of distributed parameter networks are touched upon but are soon forgotten by the student.

With the proliferation of digital devices and the expansion of software and computer technology there is little time left to emphasize the physics that underlies all electrical phenomena. Physics is usually taught in a very mathematical manner. Subjects such as radiation, antenna, and transmission line theory receive short treatment and involve Maxwell's equations, boundary value problems, and vector calculus. The student is rarely provided a connection between these topics and the real-world issues of interference control.

The subject of interference control (grounding and shielding) can be explained by using basic physics. This subject cannot be approached by using circuit theory alone. The concepts and symbols of circuit theory are invaluable because no other tools are available. But there must be an understanding that the symbols may not imply electrical components. The geometry of a structure is rarely treated in circuit theory, and yet geometry is at the heart of understanding interference processes. This is an important point that will become much clearer as the reader proceeds through the book.

Many topics cannot be treated using circuit theory. Examples might include lightning phenomena, microwave transmission, antenna radiation, noise on power distribution systems, skin effect, shielding effectivity, aperture penetration, and so forth. These processes all relate to interference, and physics provides the only tools that can be used to explain the processes. It is for this reason that the book must start out by a treatment of electromagnetic phenomena. The concepts are far more important than the ability to solve specific problems. For this reason a mathematical approach is not taken.

If the reader is patient and reads the basic material first, the entire subject of interference control (grounding and shielding) will become clearer. It is difficult to change old habits and solidly entrenched concepts. It is only through change, however, that new insight can be gained.

1.2 THE LOW-FREQUENCY SIGNAL DEFINITION

An arbitrary definition of low frequency might be signals below a frequency of 100 kHz. To someone dealing with the physics of light, a signal at

10 GHz might be considered low frequency. In electronics, interference is related to both electric and magnetic field coupling. If the frequency is low enough, magnetic field coupling can usually be ignored. There are a few exceptions where the magnetic field can cause trouble. Examples include (a) near power transformers and (b) near conductors carrying very high amperage or very low frequency transmission systems. There are many subjects that can be treated without involving the magnetic field, and this is the reason for this 100-kHz definition. We begin our discussion by considering electrostatic phenomena. There are many interference mechanisms that can be understood by treating this one area of physics.

1.3 CHARGE

The basic unit of charge is the coulomb. If charge flows in a conductor such that 1 C (coulomb) passes a given point in one second, 1 A (ampere) is said to flow. The coulomb is a large unit. It takes 6.28×10^{18} electrons to make up a charge of 1 C. The number of electrons that make up current flow in a conductor is a very small fraction of the electrons present. This is true even if the current levels are 100,000 A. The symbol Q is used to represent a charge.

1.4 FORCES BETWEEN CHARGES

Various experiments can be devised to demonstrate that charged bodies experience a force of attraction or repulsion. The forces exist between the charges themselves, although the forces appear to arise between the bodies. If an excess of electrons is placed on two bodies, they are both said to be negatively charged; so charged, the bodies will repel each other. If electrons are removed from two bodies, they are said to be positively charged. So charged, they will also repel each other. Experimentation with oppositely charged bodies shows that they will attract each other. The forces that exist are the result of the charges trapped on or in the bodies.

The electrostatic force f between two small charged bodies is proportional to the charges Q on the two bodies and inversely proportional to the square of the separation distance r. Expressed mathematically without regard to units,

ELECTRIC FIELD

$$f = \frac{Q_1 Q_2}{r^2} \tag{1}$$

If the charges are of equal sign, the positive product indicates a force of repulsion.

The force between two charged bodies varies as a function of the medium in which they are imbedded. These forces are greatest in a vacuum. The reduction in force is a measure of the dielectric constant k of the media. The force between two charged bodies immersed in a media is

$$f = \frac{Q_1 Q_2}{kr^2} \tag{2}$$

where Q_1 and Q_2 are the charges on the bodies and r is the separation distance. These forces are actually between charges; but since the charges cannot leave the body, the forces appear to be between bodies.

1.5 ELECTRIC FIELD

The interaction between charged bodies gives rise to the concept of an electric field. The field is usually represented by lines that connect between opposite charges. One line is drawn for each unit of charge. The field has a pattern that depends on how the charges are distributed. The field pattern can be constructed by using a very small test charge. The test charge must be small enough not to add to the field being measured. The force on the test charge measures the intensity of the field. The direction of the force is in the direction of the field. The lines representing the field are parallel to the direction of the force. The intensity and direction of the field at each point in space defines the field. A field with direction and intensity at each point in space is called a *vector field*. The lines representing the field are also called *lines of force* or *lines of electric flux*. The electric field or E field is a force field.

If a charge Q is placed on a conducting sphere located in a vacuum, the force field is represented by a set of straight lines that appear to radiate from the center of the sphere (see Figure 1.1). The charges on an isolated conducting sphere will distribute themselves uniformly on the surface of the sphere. The lines of force will actually terminate perpendicular to the surface of the sphere. The lines must radiate out and terminate on opposite charges which in this case are located at infinity. In the practical world these lines of force terminate on other conductors located at a great distances.

The electric field is represented by the letter E. The intensity of the E field is given by

$$E = \frac{Q_1}{kr^2} \tag{3}$$

where Q_1 is the charge on the sphere, r is the distance from the center of the sphere, and k is the dielectric constant. For this case the lines of E are these radial lines. The force on a unit test charge is given by Eq. (3).

1.6 THE E FIELD FOR CHARGE DISTRIBUTIONS

Each charged body in a system contributes to the E field independently. At each point in space the force field that results is the sum of the forces from each source. Because the direction of the force from each charge will vary, the total E field is the vector sum of the contribution from each charge. The shape of the E field in three dimensions can be quite complex and difficult to describe mathematically. In this book it is only necessary to understand the concept not to derive equations for the lines of force or field intensity at each point in space.

1.7 THE CONCEPT OF VOLTAGE

If a unit test charge is moved in an electric field, work must be done on the system. If the force over a one-centimeter span is one dyne, then the work required to move one centimeter in the direction of the force is one erg.

Without getting into systems of units, it is important to understand the definition of voltage. A voltage difference is the work required to move a unit of charge between two points. In general, if these two points are on separate conductors, then the work required to move a unit charge between these two points is the voltage difference one can measure with a voltmeter. It is important to realize that voltage is a field concept. A measure of voltage for purposes of standardization might involve a standard cell or a Josephson junction, but this is a different subject.

VOLTAGE GRADIENT

Now comes an important point. Voltage differences imply a charge distribution. This is because there can be no voltage without an electric field. In a system of conductors an electric field means there must be a charge distribution on the conductors. Circuit theory does not give this picture of charge distribution when there are voltages.

The work required to move a unit of charge through a potential difference (voltage) is analogous to lifting pails of water into a water tank. The work required per pail is a measure of the potential energy added by that pail of water. In this case, work is done in a gravitational field. One pound of water lifted 20 feet represents work of 20 foot-pounds. This added potential energy does not depend on the path taken by the pail. The same thing is true when moving a unit charge. The measured voltage does not depend on the path taken by the charge. A voltage difference in a circuit is also called a *potential difference*. This expression relates to the potential energy stored in the system by charges. In both the electric and gravitational field cases the work done on the system is independent of the path taken. Both of these fields are called *conservative fields*.

1.8 VOLTAGE GRADIENT

The steepness of a hill at a given point is a measure of its potential gradient. The steepness could be measured as the ratio of potential energy increase per unit of horizontal motion. The direction of motion should correspond to the direction of steepest ascent.

In the electric case a test charge Q is moved in an electric field E. The system undergoes a change in potential energy as the charge is moved. If the direction of motion is in the direction of the field line, this increase in potential energy will be maximum. By analogy the potential gradient can be measured by noting the change in potential energy for a unit motion in the direction of the E field. The gradient G is simply the ratio of voltage change to the unit distance moved:

$$G = \frac{\Delta V}{\Delta x} \quad (4)$$

Solving for ΔV we obtain

$$\Delta V = \text{work/unit charge} = [\text{force/unit charge}]\Delta x = G\Delta x \quad (5)$$

It is obvious that G equates to the definition of E. This means that the value of E is the steepness of the potential hill. More precisely,

$$E = \frac{dV}{dx} = \text{grad } V \tag{6}$$

where x is in the direction of maximum change.

1.9 SPHERICAL CONDUCTOR WITH A CHARGE

The simplest system to consider is the charged conducting sphere. The forces between charges causes them to spread uniformly over the surface. If there were charges inside the conductor, the forces between charges would cause them to move. Since there is no current flow, the only conclusion that can be accepted is that the charges must rest on the surface. All the forces are perpendicular to the surface of the sphere, and the charges cannot leap out into space. Any tangential E field would result in a surface current, and this cannot exist in a static environment. This means that the E field lines are all perpendicular to the surface of the sphere. If there is current flow, then the picture is obviously different.

The E field at any point outside the sphere is given by

$$E = \frac{Q}{kr^2} \tag{7}$$

where Q is the charge on the sphere, r is the distance from the center of the sphere, and k is the dielectric constant. The work required to bring a unit charge from infinity to a point r is given by the integral of force over distance:

$$W = -\int_{\infty}^{r} \frac{Q}{kr^2} \, dr = \frac{Q}{kr} \tag{8}$$

If the potential at infinity is assumed to be zero, then the potential at the surface of the sphere is

$$V = W = \frac{Q}{kr} \tag{9}$$

where R is the radius of the sphere.

1.10 CAPACITANCE

The ability of a system to store charges for a given set of voltages is described by a concept called *capacitance*. This ability to store charges is strictly a matter of conductor geometry. Charges need not be present to have a capacitance.

For the sphere just discussed, the ratio of charge to voltage on the surface is the self-capacitance of the sphere. This ratio is

$$C = \frac{Q}{V} = kR \tag{10}$$

A sphere with twice the radius can store twice the charge for the same voltage. Later when units are discussed, the capacitance of the earth can be calculated.

1.11 THE DISPLACEMENT FIELD D

It is convenient to introduce another vector field called the *displacement field*. This field depends on charge only and not on the dielectric constant. In the case of the sphere, the D field at every point outside the sphere is given by

$$D = \frac{Q}{r^2} \tag{11}$$

This implies that the relationship between D and E is

$$D = kE \tag{12}$$

When the charge distribution is known, it is desirable to consider the D field first and then calculate the E field based on the arrangement of dielectrics. When the voltages are known, it is best to solve for the E fields and then use the dielectric values to calculate the D field. Remember that the D field is continuous through dielectric boundaries unless there are charges on this boundary. This is not true of the E field where it changes abruptly at a dielectric boundary.

1.12 FIELD REPRESENTATIONS

A static electric E fields exist everywhere in the space around conductors carrying charges. The E field can be represented by lines of force that connect the various charges. These lines of force or flux lines originate on positive charges and terminate on negative charges. The number of lines leaving a charged body is proportional to the amount of charge. It is also possible to represent the field by a larger number of lines to give a useful picture of the force field. Figure 1.1 shows the radial field representation around a simple sphere.

Another representation of the static E field involves tubes of flux. The tubes must occupy all the space between conductors. For a simple sphere the tubes leaving the surface must expand as they proceed away from the surface. This expansion is a measure of the weakening field strength as the radial distance is increased.

The electric flux density is measured by considering the amount of electric flux (number of lines) that cross a unit area in space. In Figure 1.2 the tube of flux leaving the sphere at radius r_1 is the volume between four radials. This tube leaves the sphere over a surface area S_1. At a distance r_2 the surface area cut by the tube is S_2. This second area is part of a larger sphere.

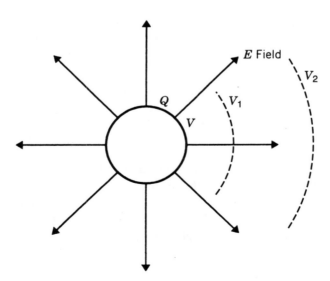

Figure 1.1 Lines of force representing the electric field E around a charged sphere.

FIELD REPRESENTATIONS

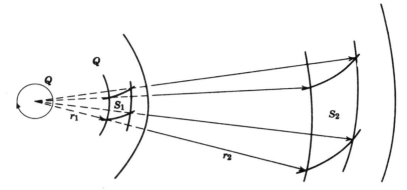

Figure 1.2 Tube of flux leaving the surface of a charged sphere.

The definition of flux that corresponds to this picture requires that

$$dN = D \, dS \qquad (13)$$

where dN is the element of flux, D is the displacement field at this same point, and dS is the element of surface area. An important theorem results when this idea is applied to the sphere in Figure 1.1. The total flux leaving the surface of the sphere can be found by integration. Since D is always normal to the surface of the sphere, the integral is

$$\int_S D \, ds = 4\pi r^2 D = N \qquad (14)$$

But we know that

$$D = \frac{Q}{r^2} \qquad (15)$$

Therefore we conclude that

$$N = 4\pi r^2 \left(\frac{Q}{r^2}\right) = 4\pi Q \qquad (16)$$

This expression states that the total flux leaving a charged sphere is $4\pi Q$.

This idea of flux leaving a region of charges can be broadened. In general, tubes of flux are not as easy to draw as the lines of force. The total flux crossing any enclosing surface depends only on the total charge inside the enclosure. Stated mathematically

$$\int_S D\, dS = 4\pi Q \tag{17}$$

This statement is known as *Gauss' electric-flux theorem*.

1.13 POINTS OF DIFFICULTY

The sphere in the previous section is convenient for a discussion of fundamental concepts without undue mathematical complexity. In other geometries the charge distributions are not uniform and the lines of force are not straight lines, but the ideas remain constant.

The sphere in Figure 1.1 cannot be reduced to having a zero radius without the E field going to infinity. In any practical arrangement, charges exist on at least microscopic volumes. The point charge idea poses no problem if the physical significance of an infinite E field is ignored. The field around a tiny sphere is indistinguishable from the field of a point charge located at the center of the sphere. It is simply a mathematical convenience to consider point charges, and nothing is lost by this assumption.

It is sometimes difficult to accept the idea of fields at infinity. The terminating charges at infinity are a mathematical necessity. In reality, all lines of force terminate on conductors. In most practical cases the earth is that end conductor.

In practice the earth is a source for all charges plus and minus. Any charge that is needed can come from the earth. Thus the earth is the termination (practical infinity) for all lines of force.

The smallest charged particle is the electron. The infinitesimal charges required in the mathematics does not exist. Even though charge distributions are actually not continuous the concepts that are derived are extremely accurate.

1.14 MKS SYSTEM OF UNITS

Gauss' flux theorem in Section 1.12 contains a factor 4π. It is desirable to select units so that this factor does not appear. With this factor, removed flux and charge are equivalent. Thus we have

MKS SYSTEM OF UNITS

$$N = \int_S D \, dS = Q \tag{18}$$

This equality can be created by changing Eq. (2) to have the factor 4π in its denominator.

In the MKS system (meters, kilograms, seconds) the charge is expressed in coulombs and the correct proportionality between units can be absorbed in an added dielectric factor ϵ. Thus we have

$$f = \frac{Q_1 Q_2}{4\pi \epsilon r^2} \tag{19}$$

and the E field is given by

$$E = \frac{Q}{4\pi \epsilon r^2} \tag{20}$$

and

$$D = \epsilon E \tag{21}$$

It then follows that the potential at a distance from a charged sphere is

$$V = \frac{Q}{4\pi \epsilon r} \tag{22}$$

and the capacitance of the sphere is

$$C = 4\pi \epsilon r \tag{23}$$

Obviously the new dielectric factor is not unity in a vacuum. If ϵ_v is the dielectric constant in a vacuum and ϵ is the dielectric constant in a media, then the relative dielectric constant k is equal to

$$k = \frac{\epsilon}{\epsilon_v} \tag{24}$$

The value of ϵ_v in the MKS system is

$$\epsilon_v = 8.85 \times 10^{-12} \text{ F/M} \tag{25}$$

where Q is in coulombs, r is in meters, f is in kilograms, and E is in volts per meter. ϵ_v is called the *capacitivity of free space*. The letter F stands for farad, the unit of capacitance.

A complete treatment of units is given in any standard text book on electrostatics. The intent here is to demonstrate basic ideas and supply the elementary equations in usable form.

The capacitance of the earth can now be calculated using Eq. (23). If $r = 6.4 \times 10^6$ m, then $C = 711$ μF. This is a static value and should not be considered the size of a filter capacitor or a circuit component.

1.15 CHARGES ON SPHERICAL SHELLS

Consider the charged sphere of radius R_1. The work required to go from infinity to the surface of the sphere is

$$V_{D_1} = \frac{Q}{4\pi\epsilon R_1} \tag{26}$$

The potential at any other radius R_2 is

$$V_{D_2} = \frac{Q}{4\pi\epsilon R_2} \tag{27}$$

From symmetry it is obvious that any imagined surface of equal potential is also a sphere. The lines of force pass through this surface unchanged.

It is interesting to replace the sphere at radius R_2 with a very thin conducting material. This surface is a conductor, and it can hold charge on either its inner or outer surface. Lines of flux that leave the first sphere must terminate on the inner surface of the added sphere. This means that the inner surface now has a charge $-Q$. If a charge Q is now added to the outer surface, the net charge on the added sphere is zero. Now lines of force radiate from this larger added sphere. The resulting external field is everywhere identical to the case when there was only one sphere.

The two spheres just discussed are shown in Figure 1.3. A charge $-Q$ is on the inner added surface and the charge Q is on the outer added surface. The potential difference between the two spheres is

$$V = V_2 - V_1 = \frac{Q}{4\pi\epsilon}\left(\frac{1}{R_2} - \frac{1}{R_1}\right) \tag{28}$$

THE EARTH PLANE

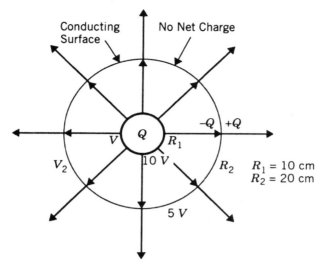

Figure 1.3 An added equipotential surface.

The capacitance of two concentric spheres is

$$C = \frac{Q}{V} = \frac{4\pi\epsilon R_1 R_{12}}{R_2 - R_1} \tag{29}$$

If a wire conductor is used to connect the outer sphere to an external large conductor such as earth, the charge on the outer surface will be drained off. There will not be an external electric field because there are no charges present on the external surface. This means that there is now no potential difference between the sphere and earth. If earth is at the zero of potential, the sphere is at this same potential. This situation is shown in Figure 1.4. The total charge on the added sphere is $-Q$. All of this charge resides on the inside surface. Initially the total charge on the added sphere was zero. After touching the wire, the total charge is $-Q$. This charge is said to be an induced charge.

1.16 THE EARTH PLANE

The conducting plane shown in Figure 1.4 often called a *ground* or *earth*. It is a large conducting body that can supply charge without changing

ELECTRIC FIELD INSIDE A GROUNDED CYLINDER

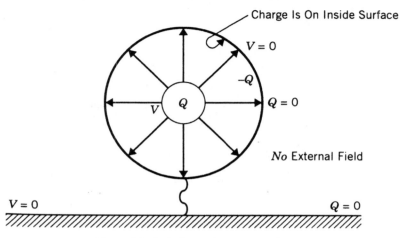

Figure 1.4 Inducing a charge on a sphere.

relative potential. The earth is actually a very complex conductor where soil conditions and contact geometry are involved. This is not an issue in the example just given. The local earth is the reference of zero potential for our discussions. If these experiments were performed in space or on an aircraft, the framework of the craft would be the local zero reference of potential.

1.17 TYPICAL CHARGE DISTRIBUTIONS

When different charges are placed on conducting bodies the field patterns can be complex. The charges move into positions that balance all the forces. There can be lines of force between the bodies and to any nearby ground plane. It should be obvious that the lines of force are always perpendicular to the conducting surfaces and that the charges are no longer uniformly distributed on the surfaces. This is shown in Figure 1.5. Note that some of the lines terminate and start on earth and that positive and negative charges can exist on the same conducting body. The potential is constant on the surface of each conductor regardless of how the charges are distributed.

The potential differences between the conductors is the work required to move a unit test charge between the conductors. The charge distribution and the voltages that result are unique. Given the voltages, the charge

CYLINDRICAL SURFACES

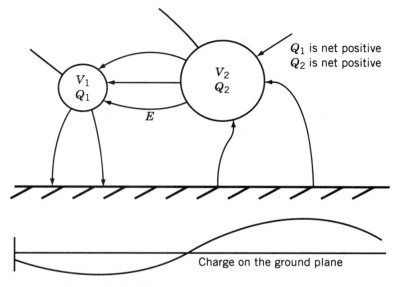

Figure 1.5 E field pattern around a group of conductors.

distribution is known; or given the charges, the voltage distribution is known.

It is important to note that the charge distribution in the ground plane will vary as the voltages change. This motion of charge is current that flows in any associated circuitry. The pattern of current flow in the earth is not simple and also involves conductors that contact the earth. The field patterns define the charge requirements and their motions.

1.18 CYLINDRICAL SURFACES

Many of the conductors used in electrical circuits are cylinders. Practical conductors include wires, printed circuit traces, and cable shields. Capacitors are made by wrapping foils separated by dielectrics into cylinders. The capacitance of two concentric cylinders can be easily calculated.

Concentric cylinders of radius r_1 and r_2 having a charge Q per unit length on the inner surfaces are shown in Figure 1.6. Gauss' flux theorem requires that the flux leaving the inner surface must equal

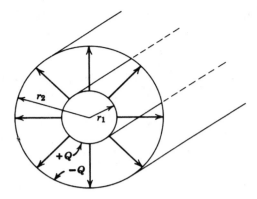

Figure 1.6 Concentric circular cylinders.

$$\int_S D \, dS = Q \tag{30}$$

where S is the surface area. At any radius r the value of D is constant. Thus at the surface of the inner cylinder the integral reduces to

$$2\pi r D = Q \tag{31}$$

Recalling that $D = \epsilon E$ the E field at any radius r is

$$E = \frac{Q}{2\pi \epsilon r} \tag{32}$$

The potential difference between the two cylinders can be derived by integrating the work required to move a unit charge between the two surfaces.

$$V_D = -\frac{Q}{2\pi\epsilon} \int_{r_1}^{r_2} \frac{dr}{r} = \frac{Q}{2\pi\epsilon} \ln \frac{r_1}{r_2} \tag{33}$$

The capacitance per unit length is given by the ratio

$$\frac{Q}{V_D} = C = \frac{2\pi\epsilon}{\ln r_2/r_1} \tag{34}$$

Unlike the sphere, this simple approach cannot be used to calculate the capacitance of a cylinder when r_2 goes to infinity.

1.19 PARALLEL PLATE CAPACITORS

Assume a charge density of Q per unit area in the geometry of Figure 1.7. The displacement field emerging from the top surface is simply Q. The electric field is given by

$$E = \frac{Q}{\epsilon} \tag{35}$$

where ϵ is the capacitivity of the dielectric between the plates. The potential difference is

$$V = \frac{Qd}{\epsilon} \tag{36}$$

The capacitance per unit area is

$$C = \frac{\epsilon}{d} \tag{37}$$

The capacitance for any area A is

$$C = \frac{A\epsilon}{d} \tag{38}$$

Two plates 1 cm apart having an area of 100 cm have a capacitance of approximately 10 pF.

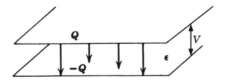

Figure 1.7 Parallel plate capacitor.

1.20 ELECTRIC FIELD ENERGY

Consider the case of a charged conducting sphere. When a unit charge is moved in its electric field, work is done. This work per unit charge is the voltage difference between infinity and the surface of the sphere. If the charge is deposited on the surface of the sphere, the total charge on the sphere is increased. This increases the field strength and the work required to bring the next unit charge is greater. This is analogous to the work added to the system when water is carried up to a tower. The tower is storing potential energy. The stored energy can be used at a later time when the water is allowed to drop in height. In the case of the charged sphere the energy stored is also potential in nature. If later this charge is allowed to drain off through a resistor, the energy will be released as heat.

Where is the potential energy stored when a charge is moved in a field? The only thing that changes is the field strength. It is thus necessary to accept the idea that the field itself must store the energy. Charge in itself does not store energy any more than a bucket of water. It is not always easy to accept ideas when they are so intangible. An electric field is not visible, but then neither is a gravitational field.

In a charged capacitor the internal field stores the energy between the plates. If the capacitor is large enough, the energy stored can be used to operate a circuit for parts of a power cycle. The energy stored in an electric field is what causes lightning. The energy stored in a gravitational field is what causes an avalanche. Both fields store potential energy.

The amount of energy stored in space can be calculated by considering the field captured inside a given volume. Such a volume might be the space between two parallel plates having a capacitance C.

If the voltage is V and a new charge ΔQ is added, the added work is $V \Delta Q$. The total work for all charges is given by integrating the work for all charges:

$$W = \int_0^Q \frac{Q\,dQ}{c} = \frac{Q^2}{2c} \tag{39}$$

Since $C = QV$, this can be rewritten as

$$W = \tfrac{1}{2} CV^2 \tag{40}$$

The value of the E field is $E = V/d$. The capacitance is given by Eq. (38). Noting that $\epsilon E = D$ the energy in terms of the capacitance geometry is

$$W = DE \cdot \text{(volume)} \tag{41}$$

Equation (41) is perfectly general and applies to all space. If a given volume of space has a fixed E field, then the energy stored in that space is given by Eq. (41).

1.21 SELF- AND MUTUAL CAPACITANCE

The capacitance of two concentric spheres and of two parallel plates has been discussed. In these cases the fields were defined by a single potential difference. In general, when there are a group of conductors a single capacitance figure is inadequate.

A self-capacitance is defined as follows: Place a voltage on a single conductor (V_1) and force all other conductors to be at zero potential (grounded). The ratio of charge on that conductor (Q_1) to the voltage applied is the self-capacitance. The symbol used is C_{11}.

A mutual capacitance is defined as follows: Place a voltage on one conductor (V_1) and force all other conductors to be at zero potential (grounded). Measure the charge induced on a second conductor (Q_2). The ratio of charge on the second to the voltage on the first conductor is a mutual capacitance. The symbol used is C_{12}. This term is also called a *leakage* or *parasitic capacitance*.

It must be recalled that a grounded conductor can have a charge on it. All that is required is that field lines terminate on that conductor.

If there are five conductors, including earth (ground), there will be four self-capacitances and six mutual capacitances. It can be shown that mutual capacitance C_{12} equals C_{21}. The mutual capacitances in this example are C_{21}, C_{31}, C_{41}, C_{32}, C_{42}, and C_{43}.

Measuring a system of mutual capacitance has its problems. The very process of grounding conductors to force a zero potential adds conductors that change the field geometry. It is very difficult to adjust all of the induced charges so that the potentials are all zero without this added grounding conductor. In most practical problems a single mutual capacitance term is under consideration and a measurement is quite practical.

It is interesting to note that an induced charge is always of opposite sign to the impressed voltage. This means that all mutual capacitances are negative. Similarly, all self-capacitances are positive.

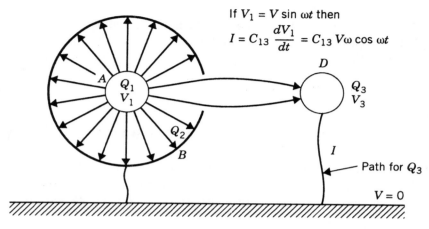

Figure 1.8 Example of mutual capacitance.

1.22 EXAMPLE OF MUTUAL CAPACITANCE

Figure 1.8 shows a shielded conductor being influenced by an external conductor carrying a signal. The shield is not perfect and some of the external field terminates on the inner conductor. The inner conductor is shown grounded. For the impedances normally encountered, the mutual capacitance is a large reactance compared to the circuit impedance Z.

The capacitance coupling shown in Figure 1.8 is very simplistic. In general, cable coupling involves several mechanisms. This topic will be covered in later sections.

In Figure 1.8 if the voltage V is varying, then the induced charge Q must also vary. This varying Q is a current. This changing current must flow in impedance Z which is a signal on the conductor of interest. The higher the frequency of the signal V, the more current will flow in the mutual capacitance and the greater the coupling.

1.23 IMPORTANCE OF ELECTRIC FIELD CONCEPTS

An electric field exists when there are voltages. The field configuration shows the location of all of the charges involved in that conductor geometry.

A POWERFUL TOOL

When the voltages change the charges must move. This motion of charge follows patterns that could not be deduced from circuit theory.

The geometry of a circuit can be arranged so that interfering electric fields can be kept out of critical regions. Conversely, the fields from a circuit can be fully contained and never allowed to escape. This viewpoint is not available using circuit theory. These concepts are broadly covered by the term *electrostatic shielding.*

When there are voltage differences, there is energy stored in fields. This energy cannot be ignored or dissipated in zero time. In many instances the decay of stored energy is interference. Two very important examples are electrostatic discharge (ESD) and lightning. Because these interference processes must be considered in most designs, they will be discussed later in more detail.

1.24 A POWERFUL TOOL

The charge distribution in Figure 1.4 gives some insight into a powerful tool. The field between the two spheres is independent of the external charge on the outer sphere. In effect the inner field is protected. In the language of interference control the inner sphere is said to be shielded. When a circuit is totally surrounded by an ideal conductor, then nothing can cross this barrier. Inner fields cannot get out and external fields cannot get in. This is a simple idea and it represents one of the very few tools that can be used to control interference.

2
MAGNETICS

2.1 INTRODUCTION

The magnetic field dominates in many interfering processes. This might be true near conductors carrying high current or near power transformers. The magnetic field dominates near a conductor carrying a lightning pulse or near an electrostatic discharge (ESD) pulse.

The concept of a magnetic field is not strange to us. We accept the fact that a compass needle lines up with the earth's magnetic field. We know the aurora borealis is caused in part by the earth's magnetic field. We use magnets for door latches and apply solenoids to move mechanical parts. Motors and transformers are all about us and make our industrial society work. We rely heavily on magnetic processes, and yet this area of physics is not well understood by many engineers.

Magnetic fields are unfortunately not as easy to approach as electric fields. The reason lies in part because there is no measurable quantity analogous to voltage. Magnetic field patterns tend to be more difficult to deal with than their electrostatic counterpart. Most magnetic processes are dynamic, while in electrostatics it is possible to deal with static charge distributions. There are static magnetic fields (permanent magnets), but these devices usually pose no interference problems.

In all interference processes, both electric and magnetic fields are involved. A practical approach that provides good insight is to calculate the interference first using the electric field and then the magnetic field. If one mechanism seems to prevail, then the other mechanism is ignored. The answer might not be accurate, but that is not the issue when the concern is interference. The idea is to reduce the interference until it no longer impacts performance. An answer that is within 20% is quite adequate. If in the design the interference is below some threshold level, the design will pass the test.

2.2 LINES OF FORCE AND FLUX

The magnetic force field can be graphically displayed by drawing lines of force or flux lines. Where these lines are close together, the field intensity is the greatest. It is obvious that the magnetic field near a current carrying conductor is more intense than the field at a distance. The lines of force can be followed by moving a small compass in the field. The compass points in the direction of the field, and the torque on the compass needle indicates the intensity. Thus each point in space has intensity and direction. These are the characteristics of a vector field.

Magnetic flux ϕ is the product of field strength and area where the area is perpendicular to the lines of force. In the case of a solenoid, each loop of wire contributes to the flux. The field lines inside a solenoid are essentially straight lines. All of the flux passes through the solenoid and crosses at right angles to the axis of the solenoid. The amount of flux is directly proportional to the current flowing in the coils of wire.

Work must be done on the system when a small loop of current is moved in a magnetic field. There is a force between the field of the current element and the field being measured. This process is philosophically correct but obviously not practical. Assuming the loop can be moved, the work that is done adds to the energy stored in the magnetic field. The equivalent of a charge in the electrostatic case would be a monopole. Unfortunately, no one has ever found one, and the idea of adding energy to the field by moving a unit current loop is strictly academic. Because of *IR* losses a sustained field cannot exist without a source of power. The only practical example of a sustained field would be around a superconductor. Here, if work were done in the magnetic field, the current would increase.

2.3 THE *B* FIELD

The magnetic field lines around a long conductor carrying current are circles. This can be seen by placing a conductor carrying a dc current through a piece of paper sprinkled with iron filings. The filings will try to line up with the lines of magnetic force.

Experiments show that the strength of the field is proportional to current and is inverse to the distance from the conductor. The magnetic field in space is called a B field or induction field, and it has intensity and direction at every point in space. The term *induction field* will become clearer when Faraday's law is discussed. B is a vector force field just like the E field in electrostatics. B thus describes the magnetic field at every point in space. The E field lines terminates on charges, but the B field lines form continuous loops in space. This type of field is said to be nonconservative. Consider the work done on the system along a line of force. Each time a loop is traversed the work done on the system is increased. This is unlike the electrostatic case where the work is a function of position independent of the path taken.

2.4 THE H FIELD

Just as in electrostatics it is convenient to introduce a second field. In electrostatics the D field was added and it resulted from charges only. In magnetics an H field is introduced that relates only to current flow. The ratio between the two fields is the permeability of the medium.

$$B = \mu H \qquad (1)$$

In a magnetic medium the H field is reduced by the permeability factor. This corresponds to the reduction of E field inside a dielectric.

Ampere's law results from experiments and states that the the line integral of H around any loop equals the current threading that loop. In the case of a long conductor, H is a constant along a circle and the integral simplifies to

$$\oint H \cdot dl = 2\pi r H = I \qquad (2)$$

This geometry is shown in Figure 2.1.

Consider the case of a solenoid with n turns. This geometry is shown in Figure 2.2. Ampere's law requires that the integral of H along any closed path through the solenoid is

$$\oint H \cdot dl = 2\pi nI \qquad (3)$$

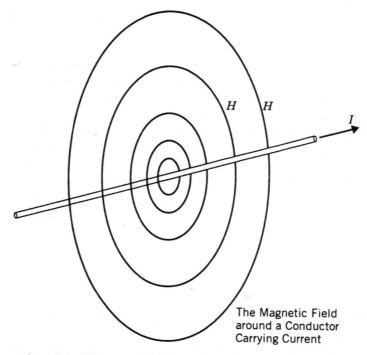

Figure 2.1 The magnetic field around a conductor carrying current.

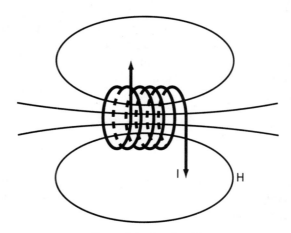

Figure 2.2 A solenoid.

THE MAGNETIC CIRCUIT

where I is the current in n turns of wire. Inside the solenoid the H field is concentrated. Outside of the solenoid the magnetic field is weaker. If the assumption is made that the field is uniform through the center of the solenoid, then the total flux is HA, where A is the cross-sectional area.

2.5 FARADAY'S LAW

When a loop of wire is moved through a magnetic field, a voltage will appear at the ends of the loop. If the loop is stationary and the amount of flux crossing that loop changes, then there will also be a voltage across the ends of the loop. The relationship between flux change and voltage is called *Faraday's law*. If there is a voltage on a loop of wire, then there must be changing flux threading that loop. Faraday's law is the basis for both motor and transformer action. The process of developing a voltage from a changing magnetic field is called *induction*.

Experiments show that the voltage across a loop of wire coupling to a B field is

$$V = A \frac{d\phi}{dt} \tag{4}$$

where $\phi = BA$ and A is the area of the loop.

If there are n turns of wire, the voltage is

$$V = nA \frac{d\phi}{dt} \tag{5}$$

2.6 THE MAGNETIC CIRCUIT

When coils of wire are placed around an iron core (permeable material) and voltages are applied to the coil, the resulting flux tends to concentrate in the iron. The reason for this concentration is simply that this flux pattern represents the smallest amount of field energy storage. Nature always runs all systems in the most efficient way possible. This flux pattern stores the least amount of field energy.

If a small air gap is provided in the iron, the nature of the field in the gap is different than the field in the iron. In the gap the H field is equal to B, while in the iron the H field is B/μ. The lines of B are continuous at an interface just as for the D field in electrostatics. The lines of H are discontinuous at an interface just as the E field is discontinuous in electrostatics. The field pattern for a gapped core is shown in Figure 2.3.

When a coil of wire of n turns is connected to a voltage source, a magnetic B field must thread the coil. The B field must satisfy the conditions stated in Eq. (5). This equation requires that current must flow to establish the flux of a changing B field. If the voltage is a sine wave, then the B field flux must also be a sine wave but shifted 90°.

The current that flows in the n turns of wire establishes the H field. This current is called the *magnetizing current*. At 60 Hz if the iron were not present, the current would have to be enormous. In the presence of a permeable material, the H field is reduced by the permeability and the current becomes a reasonable value. For example, consider a transformer at 60 Hz with a magnetizing current of 50 mA. Assume the permeability of the core is 20,000. If the core material is removed, the magnetizing current rises to 1000 A. Obviously this is totally impractical.

Going back to the magnetic path with a small gap, the voltage on the coils of wire again defines the B field. The H field in the permeable metal is B/μ, but in the gap where the permeability is one we have $H = B$. This means that the value of H in the gap is high. Using Ampere's law around the flux path we obtain

$$Hl_1 + Hl_2 = nI \tag{6}$$

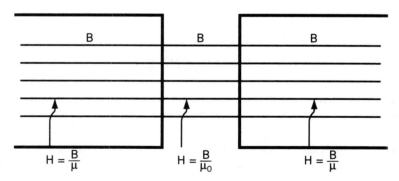

The B field is continuous in the gap

Figure 2.3 A gapped magnetic circuit.

THE TRANSFORMER

With the added term Hl_1 the current requirement is now higher. In effect, the permeability of the magnetic circuit has been reduced. Equation (6) looks very much like the sum of voltages around a circuit loop. The term nI is a magnetizing force (voltage), and each term Hl represents a voltage drop (IR). This concept can be carried further by discussing a parameter called *reluctance*. Reluctance is the magnetic equivalent of resistance in an electrical circuit.

Magnetic recording heads are designed with a controlled gap. The gap must allow the magnetic field to penetrate the recording media over a small area. Head design and manufacturing methods are carefully protected. The coils are so small that they must be wound using microscopes in a clean room environment.

Magnetoresistive materials can be used in magnetic heads to read data stored on magnetic media. The resistance varies as a function of field strength. A current flowing in the material is modulated by the magnetic field and sensed as a voltage. The heads are by necessity very small and the signals that result must be amplified. Care must be taken to avoid interference from any nearby magnetic fields. The treatment of the input cable follows the guidelines given in later chapters.

2.7 THE TRANSFORMER

When turns of wire are placed around a magnetic path, the voltage applied to the coil establishes a B field in the iron. When a separate or secondary coil is added around the magnetic path, the voltage across this coil is also coupled to the changing B field by Eq. (5). A voltage on this added coil will ideally follow the wave form of the applied voltage. The voltage will be proportional to the number of turns. A square wave of voltage on the primary causes a square wave of voltage on the secondary and this voltage is without sag. This assumes that the transformer iron is not saturated.

When a load is applied to the secondary coil, current must flow in this coil by Ohm's law. This current cannot change the magnetic parameters that are defined by the primary voltage. To maintain the correct B field a balancing current or countercurrent must flow in the primary coil. In other words, the load ampere turns of the secondary coil must be balanced by an equal number of ampere turns in the primary coil. If the primary has 1000 turns and the secondary has 100 turns, the load current in the primary coil will be one-tenth the load current flowing in the secondary coil. The total primary current is now the magnetizing current plus the load currents

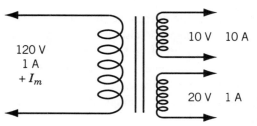

Figure 2.4 A simple transformer showing ampere-turn balance.

needed to balance the ampere turns of all secondary loads. This current balance is shown in Figure 2.4.

When a step voltage is applied to a transformer coil, the B field must increase linearly. When the B field is at its upper limit, the core is said to be saturated. When the core saturates, the magnetizing current rises abruptly. In most applications the voltage is reversed in sign before saturation occurs, and the B field then reverses its direction of travel. The flux passes through zero and eventually the core is again saturated in the opposite direction. When sine waves are applied to a transformer primary the B flux pattern is a cosine wave. The maximum B field occurs at the zero crossing of volta e. The voltage and B field pattern are shown in Figure 2.5.

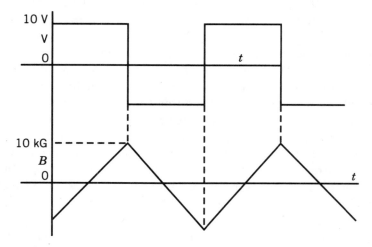

Figure 2.5 The voltage and field patterns in a typical transformer.

2.8 HYSTERESIS—MAGNETIC MATERIALS

The relationship between B and H for various materials is important in the design of transformers. At 60 Hz the iron that is used is different from than that used in switching regulators. The field in and around the core can be a source of interference. The problems at 60 Hz are considerably different than the problems at 100 kHz. At high frequencies the number of turns required to support the flux is reduced. This is in the direction of higher magnetizing current. Fortunately at high frequencies the value of B can be kept small, and this is in the direction of limiting the magnetizing current. In any practical design a compromise solution must be found. At high frequencies any leakage inductance plays an important role. This inductance adds a series reactance which limits power transfer. Careful attention must be paid to winding geometries to limit any leakage flux. Close coupling unfortunately adds to the parasitic capacitance between coils. This problem is addressed in later chapters.

All magnetic materials exhibit hysteresis. Hysteresis is broadly the relationship between the B and H fields in the iron. Manufacturers usually give curves showing the B/H relationship for sinusoidal B excitation. The relationship is nonlinear and varies with peak values of B, with core material, and with frequency. For cores used at frequencies above 20 kHz the B/H relationships may be given for square waves.

Permeability is the ratio of B to H. Since the hysteresis curves are nonlinear, a simple number cannot describe the relationship. The number quoted is often the ratio of peak values. This tends to exaggerate the useable permeability. A typical B/H relationship is shown in Figure 2.6. Curves of permeability versus peak B can be quite helpful. Manufacturers also provide a value of initial permeability. This is the ratio of B to H for low values of B. This gives the designer some idea of performance when the flux levels are low (small signals). This specification is usually listed when the material is intended for low flux applications.

If the resistance of the primary coil in a transformer is low, the B field follows the voltage waveform. The magnetizing current waveform will be nonlinear, but this will not affect the voltage wave shape as seen on the secondary. In small power transformers the primary resistance can be quite high. The nonlinear magnetizing current flowing in the primary resistance modifies the voltage generating the B field. This results in secondary voltages that are distorted. The distortion usually makes the waveform look somewhat square. This pattern is sometimes seen as interference. The problem is even more complicated when load currents flow over a fraction

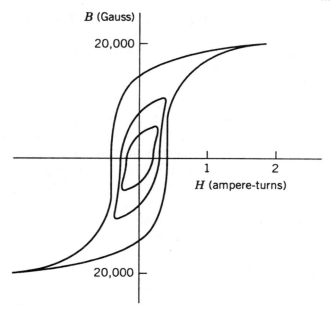

Figure 2.6 A typical hysteresis pattern.

of the power cycle. This occurs when capacitor input filters are used in rectification. Between magnetizing current effects and nonlinear loads, the waveforms on the secondary can be greatly modified.

The air gap discussed earlier can be distributed throughout the core material. This is the case when ferrites are used. In a ferrite core, the magnetic domains are extremely small and the domains are separated by a ceramic filler. This makes the core material nonconductive, and this reduces eddy current losses to acceptable levels at high frequencies. The permeability that is listed includes the distributed gap.

Power and audio transformers are often built using stacked laminations. These laminations reduce eddy current to an acceptable level. At 60 Hz, typical laminations are 0.016 inches thick. At 400 Hz the laminations are often 0.006 inches thick. Thinner laminations are usually impractical. One-mil strips of metal can be wound into a torroid to form a core. Unfortunately the coil winding procedures are more complex and the cores are more expensive. Ferrite cores in torroidal form come in two pieces. When assembled, the two halves contact each other across a very well polished gap. The core is assembled around a bobbin containing the coils. When assem-

THE INDUCTANCE OF A SOLENOID

bled, the resulting mechanical gap is essentially zero. Cores of this type are available with permeabilities up to 10,000 and can operate at frequencies above 10 MHz. Versions are available with variable gaps so the assembly can be used to store magnetic field energy. This allows some dc current to flow in the coils without core saturation or for the component to be used as an inductor.

2.9 INDUCTANCE

In electrostatics, energy was stored in an electric field and this ability to store energy was called *capacitance*. In the magnetic case the ability to store energy in the magnetic field is called *inductance*. These concepts are geometric in nature and do not require voltage or current.

Self-inductance is the proportionality between voltage and the rate of change of current. Stated mathematically,

$$V = L \frac{di}{dt} \tag{7}$$

The inductance can be calculated directly from circuit geometry, but only a few cases lend themselves to a simple derivation. Two examples where the geometry has a simple solution are the solenoid and the coaxial cable.

Mutual inductance is defined as the voltage induced in a first circuit from current changes in a second circuit. The nomenclature is L_{12}. It turns out that the value of L_{12} equals L_{21}. The induced voltage in a second circuit can be of either polarity.

2.10 THE INDUCTANCE OF A SOLENOID

The H field inside a solenoid is given by Eq. (3). The B field is simply

$$B = 2\pi \, IN \, \mu_0 \tag{8}$$

The flux is simply the B field times the cross-sectional area, or

$$\phi = \pi \, n^2 B \tag{9}$$

This flux threads through all N turns so that the induced voltage is proportional again to N. The inductance is the flux per unit current, or

$$L = \frac{n\phi}{i} = 4\pi^2 n^2 r^2 l\mu_0 \tag{10}$$

where n represents the turns per unit length and l is the length. This equation is correct in form, but further treatment is necessary before the units all agree.

2.11 ENERGY IN THE MAGNETIC FIELD

Consider a small volume element in space where magnetic flux crosses between two parallel faces. When a unit current loop moves in this flux the work done is Bx, where x is the distance between the faces. The increment of current increases the flux, and the next current loop that moves across requires more work. This is analogous to moving charges in an E field and accumulating these charges on the plates of a small capacitor. The accumulated current loops (flux) is proportional to H, and the force is proportional to B. The flux is simply H times the area of a face. The total work done on the system is thus

$$BHAx = \frac{B^2 v}{\mu} = \mu H^2 v \tag{11}$$

where v is the volume of the element. This equation turns out to be correct for any volume element in space. This equation holds even if the field is separated from any circuit (radiating). Just as in the electrostatic case, the permeability factor needs to be adjusted to make the units come out correctly.

2.12 MAGNETIC UNITS

When Ampere's law is applied to a long conductor the H field is given by

$$H = \frac{I}{2\pi r} \tag{12}$$

LEAKAGE INDUCTANCE

In the MKS system (meters, kilograms, seconds) it is obvious that H has units of amperes per meter. When voltage induction is considered, the B field must accommodate Faraday's law. In a vacuum $B = \mu_0 H$, where μ_0 the permeability of free space is

$$\mu_0 = 4\pi \times 10^{-7} \text{ teslas meters/ampere} \tag{13}$$

The unit of B is the tesla. One tesla is equal to 10,000 gauss. (This factor is the number of square centimeters in one square meter.) The B field specifications for magnetic materials are usually given in gauss. The H field is often given in units of ampere-turns per centimeter.

To calculate the voltage induced in a circuit the H field is first calculated using Ampere's law. The B field is then calculated in teslas using $B = \mu H$. The flux involved is found by multiplying B times the area involved in square meters. The induced voltage is simply the rate of change of this flux. If more than one loop of wire is involved then the voltage is multiplied by n.

Here is an example to illustrate induced voltage. A lightning pulse of 100,000 A flows in a steel girder in a building. The pulse rises to its peak value in 0.5 μsec. The H field 1 meter away from the girder is $I/2\pi r$ or 15,915 amperes per meter. The B field is found by multiplying H by $\mu_0 = 4\pi \times 10^{-7}$. The B field is 0.02 teslas. If the loop area of a nearby circuit is 10×10 cm^2 (0.01 m^2), the flux is B × A or 2×10^{-4} lines. The voltage is simply this flux change in 0.5 μsec, or 400 V. It is easy to see why it is dangerous to place electronic equipment next to steel girders. During a lightning storm the equipment can be destroyed. The lightning current does not need to flow in the equipment to do damage.

The permeability of free space is given by Eq. (13). The permeability of a magnetic material is given by the product of μ_0 and μ, where μ is the relative permeability. The relative permeability has no units. For transformer iron the relative permeability can be as high as 90,000.

2.13 LEAKAGE INDUCTANCE

In an ideal transformer the flux developed by the voltage on the primary coil follows the core. There is always some flux around the conductors that does not follow this path. Some of the flux associated with load currents remains around the individual turns. The flux that does not thread both the primary and secondary coils is called *leakage flux*. This flux is not available to provide transformer action. It can provide interference, however.

In large power transformers the load currents can exceed 1000 A. The flux from this load current results in a large leakage flux that can impact nearby circuitry. In one installation I visited, the main distribution transformer was located on the 10th floor. Computer monitors on the 11th floor were impacted. The images had wavy lines. The distances were over 30 ft. The secondary currents exceeded 1000 A.

Large power transformers are difficult to mount. If there are nearby conductive paths (loops of structural steel), leakage flux can induce significant current to flow. This current may find its way into the entire steel structure of a building. It is clear that conductive loops in the vicinity of large distribution transformers should be avoided. Holes in transformer iron should not be used to mount the transformer because the bolts form a shorted turn around parts of the core. When insulating shoulder washers are used, a shorted turn can be avoided.

The equivalent circuit diagram of a transformer is shown in Figure 2.7. If the transformer were ideal, the magnetizing inductance would be infinite and the leakage inductances would be zero. Small power transformers have leakage inductances of a few millihenries, while magnetizing inductances often exceed one henry.

The maximum B field in a power transformer is a function of the iron that is used and not of the power rating. The maximum B field is 15 kg whether the transformer is rated 100 W or 10 kW The external fields or leakage fields can be large simply because of the large secondary currents that are possible. The fields are more extensive around a large transformer because of the geometry of the transformer.

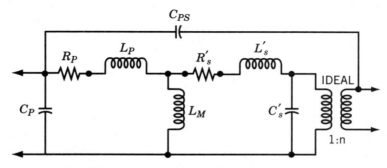

Figure 2.7 Transformer equivalent circuit. Key: Primes indicate correction factor n^2; $R'_s = R_s/n^2$. It is convenient to refer all impedance to one side of the transformer.

The large magnetizing inductance of a power transformer does not imply that the transformer can be used as an inductor to store magnetic field energy. In fact a transformer is specifically designed not to store field energy, and a small dc current will saturate the core. Energy storage is proportional to both B and H. Without a gap, H is very small along the entire magnetic path. An inductor that is intended to store field energy usually has a gap. Field energy is stored in space, and the gap is that space.

Magnetizing inductance can be measured by noting the primary current when all loads are removed from the transformer secondaries. Leakage inductance can be measured by noting the current flow when all secondaries are shorted out. The resistance of the coils can be measured with an ohmmeter.

2.14 THE AUDIO TRANSFORMER

With the advent of power transistors and off-line switching power supplies, fewer audio transformers are being used. Many designers still like the sound quality afforded by a transformer. Therefore it is still worthwhile discussing the audio transformer problem. The maximum flux density occurs at the lowest frequency of interest. In a general application some margin must be provided because the signals are not sinusoidal. At high frequencies the flux density is very low and an increase in magnetizing current can be tolerated.

An output signal transformer must often couple power. The coupling of power means that the leakage inductance and coil resistances must be considered in transmission efficiency. The transformer designer will usually specify source and load impedances that optimize the transformer's bandwidth. In some circuits the output stages are not perfectly balanced and some dc current will flow in the primary coils. In this application the transformer core should be gapped to avoid core saturation.

2.15 MAGNETIC STORMS

Sunspot activity creates enormous magnetic fields. These fields change rather slowly, but they can still induce large currents in the earth. These currents in the earth can cause problems. For example, earth currents can impact power transmission. Earth current can be high enough to be sensed by ground fault interrupters, and this results in load shedding. If not properly controlled, the power grid can become unstable and close down.

In the tundra, where steel pipes carry oil, conductive loops are formed by the steel in supporting pillars and the conductive earth. In the winter, when the frozen earth is nonconductive, the loop areas are very large. During sunspot activity, currents as large as 400 A have been observed.

2.16 SPACECRAFT POTENTIALS

As an orbiting spacecraft moves, it crosses the magnetic field of the earth. The earth's magnetic field is weak, but the velocity of the spacecraft is high. The flux crossing a conductive surface causes current to flow in that surface. This is very much like the eddy current phenomena found in transformers. The field that couples to a circuit loop will add a dc voltage. If the loop area is kept small, this voltage will also be small.

2.17 THE E AND H FIELDS TOGETHER

In the electrostatic case when voltages change, charges must move. A moving charge is a current, and a current requires an associated H field. In the magnetic field case when a current changes, the magnetic field must change. This changing field has a voltage associated with it that can be measured with any available loop of wire. The presence of a voltage in space implies an E field.

A changing E field implies an H field, and a changing H field implies an E field. These fields are bound together so that both are required anytime there is a change. This relationship is expressed in Maxwell's equations. These equations are eloquent statements that describe all electrical behavior. Fortunately, it is not necessary to use these equations to develop an understanding of the principles involved.

The important philosophical point to note here is that all electrical activity, whether it be power transfer or signal transmission, requires both a magnetic and an electrical field. The fields vary in importance as a function of power level and of frequency. In most low-frequency applications, electric field phenomena dominates. Above 1 MHz the two fields must be considered together. There is no magic boundary where electrical phenomena is dominated by only one field.

2.18 FIELDS AND COMPONENTS

All components require fields to function. This a viewpoint not often stated in circuit theory. A simple resistor requires a voltage across its terminals. This means that an E field must exist through the body of the resistor. It is this E field internal to the resistor that causes the charges to move.

Capacitors store electric field energy and release this energy back into the circuit. Inductors store magnetic field energy and return this energy back into the circuit. Semiconductors function as a result of internal fields. Note the name *field effect transistors*. Transformers function through the mechanisms of a changing B field. All electrical components require fields. It will be seen later that the transport of power takes place in fields.

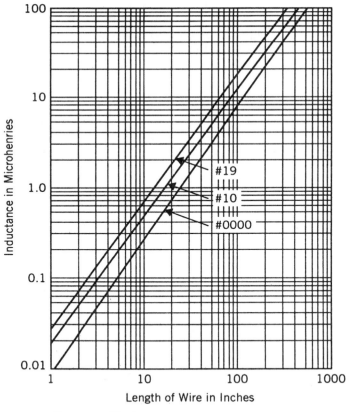

Figure 2.8 The inductance of round copper wires.

Everything is field-operated. It is important to understand that the purpose of conductors in circuitry is to transport the correct fields to all the components. It is through this viewpoint that the problems of interference will become understandable.

2.19 THE INDUCTANCE OF ISOLATED CONDUCTORS

The inductance of a single isolated conductor is given in the textbooks. This inductance is derived assuming that the conductor is a part of a large loop. A plot of inductance for various-diameter conductors as a function of length is given in Figure 2.8. Note that at 1 MHz the inductance is about the same for a #19 or for a #0000 conductor. Skin effect is included in these curves, but it is a small part of the impedance and can usually be neglected.

To bring an important point home, a #10 conductor 100 in. long has a reactance of 62.8 Ω at 1 MHz. A thin sheet of metal 100 in. square and 0.1 mm thick has an impedance of 355 $\mu\Omega$. This is five orders of magnitude lower. Brute force does not work. At high frequencies the key is always geometry. The same copper in a different shape is remarkably different.

3

THE TRANSPORT OF SIGNALS AND POWER

3.1 INTRODUCTION

This chapter describes how power and signal (information) are transported by fields. The idea that conductors carry power or carry signal is at the heart of many misconceptions. The reason for conductors is obvious. They act as rail lines to direct the flow of energy or signal. Without a set of rails the field energy cannot be guided to its destination. At microwave frequencies the rail can take the form of a hollow conducting cylinder. With lasers the energy can be well-directed as a narrow beam. Information can be transported using radiated energy (radio and TV), but this radiation is not used to transport useable power. In these applications the information is thrown out over a vast space so that many users can have access to the information.

Nature plays a mean game. Every pair of conductors can serve as a rail line for carrying signal or energy. One of the conductors can be the earth or the sides of a rack. The conductor pair could be two shields. Field energy that couples to a pair of conductors goes in both directions. Why does energy follow these conductors? The answer is simple. It is easier to follow this path than to continue in free space. Nature runs downhill as fast as she can go. It is the same principle that applies to water. Water runs downhill. It not only runs downhill, it goes to the bottom of the valley as quickly as possible and follows the river. If it could get any lower, it would.

When we arrange a circuit expecting to move signals or energy, nature goes along for the ride. Field energy can be captured by the conductor geometry and can travel the rails in both directions. Often this field energy is the interference we do not want to see. Conductor geometries in facilities are never simple. They are installed to meet needs that are often nonelectrical. Some of these conductor geometries involve power support, building steel, natural gas, telephone lines, shields, earth, racks, and conduit. Only a few of the conductors involve the signals of interest. It is impossible to totally separate the conductors of interest from all of the others. Nature is oblivious to our wishes and needs. If there are paths to take, they will be taken.

To understand how these "rails" work, it is necessary to start with simple geometries. Even with simple geometries, there is much to say. These special cases give us the insight needed to approach the more difficult problems. Again there is no need to find exact solutions. The idea is to understand the broad picture. The simple transmission line is a good place to start. The coupling of interference to transmission lines is discussed later.

3.2 THE TRANSMISSION LINE

The classical approach to transmission lines involves the idea of distributed parameters. The conductive lines are assumed to have a distributed series inductance and a distributed shunt capacitance. There is inductance and capacitance per unit length all along the line. In the general case there are also distributed series and shunt resistances to represent losses on the line. Differential equations can be developed to treat this distributed circuit. The solutions to these equations show how signals propagate down the transmission line. This approach can be used to describe several geometries that include pairs of conductors, conductors over a ground plane, and coaxial cable. It is convenient to start with a pair of conductors and omit the series and shunt resistances.

The easiest way to see how a transmission line works is to consider what happens when a voltage is switched onto the line as in Figure 3.1.

When the switch closes, charge starts to flow in the capacitance through the series inductance. Energy cannot suddenly appear in an electric or magnetic field because this would require infinite power. This means it takes a finite time for the current to flow in the first inductance and for a voltage to appear across the first capacitance. In a next short period of time the next increment of inductance and capacitance can store energy.

TRANSMISSION LINE TERMINATIONS

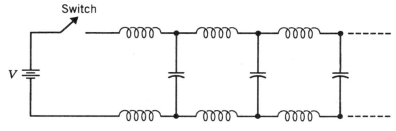

Figure 3.1 A simple transmission line.

The process of energy flow continues with time. The capacitances along the line are charged at a fixed rate, and this results in a steady current flow from the battery. The current stores energy in each element of inductance as the capacitances are charged. The result is a step wave that propagates down the line. If there were no losses or radiation, the wave would continue down the line at a fixed velocity. A person monitoring the line could watch the step function proceed like a train following the tracks.

Viewed from the voltage source, the current that flows is a steady value. With a fixed voltage and a fixed current the transmission line looks like a simple resistor R. The power flowing into the transmission line is equal to VI. Energy must actually be moving along the line as it is being supplied at a steady rate from the battery. The transmission line is functioning like a pipeline carrying energy. Just like a pipe line, the transmission line must end somewhere. If this ideal line were infinite in length, the battery would continue to supply a steady current.

If the voltage source changes value, a different amount of energy flows into the line. The new voltage waveform appears along the line delayed in time. Again all of this assumes a line without losses or radiation. For most practical transmission lines the velocity of the wave is about one-half the speed of light. Typical input resistances might vary from 30 to 600 Ω depending on conductor spacing.

3.3 TRANSMISSION LINE TERMINATIONS

Assume that the transmission line in Figure 3.1 is terminated in a resistor R. When the wave reaches the termination the current flow will still be I. The voltage source continues to supply energy, but now it flows down the line and into the resistor. The source knows nothing about this termination and continues to supply energy at the same rate.

Transmission line theory was initially developed to provide engineering tools for the radio industry. The interest was in handling carrier sine waves that were to be radiated from an antenna. This theory allows for varying line lengths and complex source and terminating impedances. It became accepted practice to say that a transmission line has a characteristic impedance. This impedance is given as a single number in spite of the fact that the term *impedance* implies a real and imaginary part. To be conventional the term *characteristic impedance* will be used in further discussions. A 50-Ω line means that when the line is terminated in 50 Ω (the characteristic impedance) the source sees the line as being infinite.

When a transmission line is terminated in a short circuit a reflection occurs when the step wave reaches the end of the line. It is impossible to dissipate energy in a short circuit. The energy that is flowing down the transmission line must go someplace. The only possibility is for the energy to turn around and go back. This requires a wave of opposite voltage and equal current. This second wave cancels the voltage at the termination. Until this wave returns to the source, the source continues to pour constant energy into the line.

At the source the voltage cannot be zero, so a third wave is sent forward to reestablish the correct voltage. This time the current must support the initial wave, the reflected wave, and the new third wave. This requires the current to be three times its initial value. Three waves are now using the same transmission line concurrently. At each reflection the current builds up by one increment of current. This is the ideal mechanism of how a short circuit develops. In practice there are many loss mechanisms that prevail, and this picture must be somewhat modified. The voltage sources are never zero impedance, and furthermore there are always current limits or some other failure mode to limit the current.

When a transmission line is terminated in an open circuit, the energy coming down the line must still reflect. The energy cannot spurt out into free space. To satisfy the boundary conditions at the termination, a wave must be returned that cancels the current. This wave when superposed on the first wave doubles the voltage. When this second wave reaches the source, the voltage is incorrect and a third wave must be sent forward that establishes the correct voltage. When this third wave reaches the termination, the current must again be set to zero. The process is repeated over and over, and waves continue to echo back and forth. This time there is no buildup of voltage. In any practical situation the source impedance

TRANSMISSION LINE FIELDS

is not zero and there are losses along the line. These reflections can be observed if the test equipment is adequate. After a few reflections the waveforms will distort and attenuate.

The reason for looking at an ideal transmission line and its reflections is to get insight into what might be happening on a group of random conductors when energy is captured. In general, few conductors run exactly parallel or are terminated. Many are terminated in short circuits and many are open-ended. Energy is not lost at reflections but in IR losses and possible radiation. The nature of any field pattern is very complicated, but obviously nature solves the problem.

3.4 TRANSMISSION LINE FIELDS

The electric and magnetic field patterns for a pair of conductors is shown in Figure 3.2. The picture assumes that the fields are fixed and unchanging after the wave has passed. The E field terminates on the conductors, and the H field surrounds the conductors. In reality the picture is far more complex. The fields are not bounded in any way, and therefore the fields must extend far into space. They cannot proceed at infinite speed, and therefore the picture that is indicated is rather crude. Fortunately a transmission line is rather efficient and little energy is actually radiated. On a printed circuit board there may be a thousand transmission lines, and because of this multiplier the radiation can be of concern.

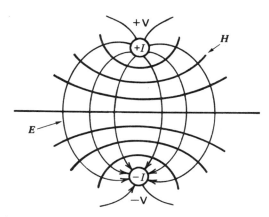

Figure 3.2 The field patterns around a transmission line.

A transmission line geometry that occurs quite frequently is the single conductor over a ground plane. The ground plane is the second or return conductor. Examples of this geometry might be traces over a ground plane on a printed circuit board, a cable running parallel to a rack wall, cables running in a metal tray, or power lines over the earth. The field configuration is shown in Figure 3.3. This pattern is simply one-half of the pattern illustrated in Figure 3.2. The current flow is the same, but the voltage is one-half. The characteristic impedance is thus about one-half the two-conductor configuration.

The characteristic impedance $Z(\Omega)$ of parallel round conductors is given in Table 3.1. Note that the characteristic impedance is a function of ratios and not of absolute dimensions. Values are given for two parallel conductors and a conductor over a ground plane. The reader may want to refer to a textbook for the characteristic impedance of rectangular conductors.

The current flow pattern in the ground plane can be seen by noting where the E field lines terminate. The current density is highest directly under the conductor. The closer the conductor is to the ground plane, the greater this concentration of current. Conductive surfaces that are not in the current path cannot reduce the impedance of this path. This fact would not make sense unless the electric field pattern were considered. Circuit theory would imply that the entire ground plane is used to return current, and this is not the case. This concentration of current occurs at all frequencies, including dc.

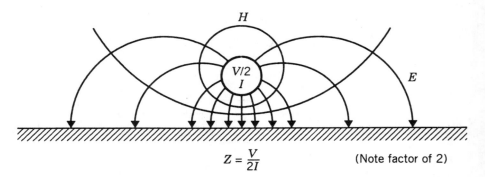

Figure 3.3 The field pattern of a ground plane transmission line.

GROUND PLANES

TABLE 3.1
Characteristic Impedance Z (Ω) of Parallel Round Conductors

L/l	Z (Ω)	H/h	Z (Ω)
1.1	53	0.6	37
1.5	115	1.0	79
2.0	158	2.0	124
2.5	188	2.5	138
3.0	212	3.0	149
4.0	248	4.0	166
5.0	275	5.0	180
10.0	359	10.0	221
30.0	491	30.0	287
100.0	636	100.0	359

3.5 GROUND PLANES

The term *ground plane* implies a relatively large conductive surface. For a power transmission line, the earth would be a ground plane; and for a printed circuit board, copper plating on the board would be a ground plane. The word *plane* is misleading because many ground planes are simply not flat surfaces. Ground planes can follow floors and walls in a room. The surface of a metal aircraft is a ground plane. The sides of a metal rack form a ground plane.

The earth as a ground plane is very complex. The resistivity depends on soil type and moisture. Quite often the soil is stratified, making the problem even more difficult. In the case of lava, granite, or desert sand a local ground plane (earth) may not exist. A ground plane plays no role in the transmission process unless field lines terminate on that plane.

The earth is the return path for lightning. Conductors that enter the earth where there is no conduction are obviously ineffective. The National Electrical Code (NEC) requires earth connections of 25 ohms or less. If this resistance cannot be met, then two connections are required. The code does not require more than two connections. The earth connection is discussed further in Section 9.6.

In some installations the ground plane is simply the racks housing the equipment. The quality of the plane depends on the bonding between the racks. Often cables are routed on the floor through the racks. The plane can be improved by placing a metal strip on the floor that threads through all racks. The author has seen cases where there are equipment problems. Once a week the bolts connecting the racks together must be retightened to clear up the problem. More will be said about bonding impedances later.

3.6 COAXIAL TRANSMISSION LINES

The two conductors involved in transmission can be coaxial There is a center conductor and a return outer conduit. This outer conductor can be solid metal or a metallic braid. The field lines for any coaxial arrangement are entirely contained. In effect the ground plane is wrapped around the center conductor. Ideally there is no external field. There are mechanisms for fields to enter or leave through the cable walls, and this is discussed later.

The containment of field implies that all the current flowing down the center conductor returns on the outer conductor or sheath. If the return current flows in an external path, then there is field external to the coaxial path and the coax is ineffective. The simplest field line configuration is shown in Figure 3.4.

3.7 SINE WAVES AND TRANSMISSION LINES

The step function discussed earlier provides one view of how a transmission line behaves. The classical approach involves sine waves and the concepts of circuit theory. With sine waves the term *impedance* can be used. The input impedance of a properly terminated transmission line is its characteristic impedance. If the line is terminated in a short or open circuit, the input impedance will be a reactance. It cannot be a resistance because there is no way for energy to be dissipated. The reactance will depend on the length of the line and the frequency.

The length of a line is often measured in units of wavelength. One wavelength corresponds to the distance traveled by a signal in one cycle. Assume the wavelength at 1 MHz is 150 m. Consider a line that is one-

SINE WAVES AND TRANSMISSION LINES

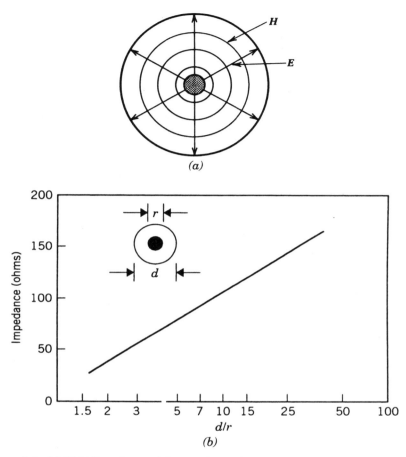

Figure 3.4 (a) Field lines for coaxial transmission lines. (b) Characteristic impedance for coaxial transmission lines.

half wavelength long or 75 m. At the input the reflected wave is equal in amplitude to the sending wave. In this case the input impedance looks infinite. If the cable is one-quarter wavelength long, the return signal is 180° out of phase and the input impedance looks like a short circuit. At intermediate lengths the cable appears to be a reactance. At low frequencies the cable looks like a capacitance.

A line that is terminated in a short circuit can appear to be an open circuit when the reflected wave is in phase. Again if the reflected wave is 180° out of phase, the line will look like a short circuit. At intermediate

52   **THE TRANSPORT OF SIGNALS AND POWER**

frequencies the line will look reactive because no energy can be dissipated in a short circuit. At low frequencies the line will look like an inductance. The input impedance of an ideal transmission line for short or open terminations is shown in Figure 3.5.

Consider a transmission line that is formed between a shielded cable and a metal panel (ground plane). Fields can be transported between the shield and panel. When the shield bonds to a conductive bulkhead, this transmission line is terminated in a short circuit. When energy using this path reaches the termination, it is reflected. When the interference is of a sinusoidal nature (carrier signal), then a steady-state phenomena may exist. The voltage level of the carrier is a function of where the measurement is made. At the termination the current will be maximum and the voltage will be zero.

When a carrier-type interference signal is coupled into a conductor geometry, the signals that appear will depend on many factors. In some geometries the refections are supportive and the interference will appear amplified. It is often stated that there are resonances present. Strictly speak-

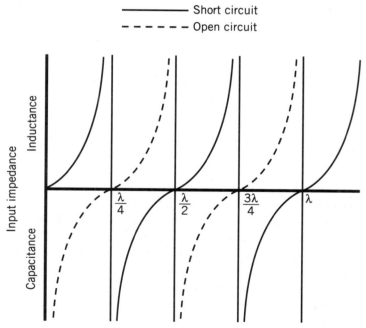

Figure 3.5 The input impedance of shorted and open terminations for a transmission line.

ing, a resonant circuit involves energy that moves back and forth between an inductance and a capacitance. In a resonant situation, energy is added per cycle until there is a balance with the energy lost per cycle. In most conductor geometries the Q of any resonant circuit would be quite low. Trying to describe coupling phenomena in terms of these resonances can be misleading.

3.8 TERMINATIONS

In instrumentation, signal cables are rarely terminated. The reason lies in the inaccuracies that result. A terminated line causes a loss of signal related to source impedance and line resistance. The line resistance varies with temperature, and the resulting signal attenuation is not easily accommodated. Even if the cable is unterminated, the amplitude versus frequency response is a function of cable type and cable length. For cable runs of more than 300 ft the response should be checked if bandwidth is important. It is possible to terminate the cable with a series RC circuit. This serves as a termination at higher signal frequencies but causes no error at low frequencies. This is sometimes called a *fractional termination*. This termination can be determined empirically using square waves. The values of RC are adjusted until the square wave response is flat for square waves at 300 Hz and 3000 Hz. For most cables the series inductance and shunt capacitance has the effect of peaking the signal at high frequencies. If the cable exhibits a loss at high frequencies, then a fractional termination is of no use.

When interference is coupled to a cable, it travels in both directions. When it reaches the signal source end of the cable, a reflection occurs unless the source impedance is matched to the characteristic impedance of the cable. When it reaches the signal termination end, it reflects unless the cable is terminated in the characteristic impedance of the cable. In most applications a voltage source drives the cable. This is a zero-impedance source except for any small inductance that might exist in an active circuit. Building up the source impedance by adding a series resistor is not advised unless it is necessary to absorb unwanted energy being returned to the source.

Matching generator impedances to load impedances is necessary in order to optimize power transfer. If power is not the consideration, then terminations or matching impedances may not be necessary. When video patterns

3.9 POYNTING'S VECTOR

All electrical energy is transported in fields. The fields have direction and intensity at every point in space. In the transmission lines the field lines are shown in Figures 3.2, 3.3, and 3.4. In every case the E and H fields are perpendicular to each other and perpendicular to the direction of power flow. It can be shown that the power density at any point in space is simply the product of E and H. If this product is summed over the entire area containing the fields, the result will be the power crossing that area. This should not be surprising in that the H field is proportional to current and the E field is proportional to voltage. Stated mathematically, the power per unit area at a point in space is equal to

$$P = E \times H \tag{1}$$

The power crossing any area is simply

$$W = \int E \times H \cdot dA \tag{2}$$

where E and H are perpendicular to each other and to the element of area.

Poynting's vector applied to a simple transmission line is shown in Figure 3.6. Poynting's vector works at all frequencies. The energy supplied from a car battery to operate headlights flows in a field. Poynting's vector applies in this dc case. The conductors direct the energy to the lamps. Without this understanding, many of the interference processes in an automobile are dark mysteries. If the return path for the current is a second parallel conductor, then the field pattern will be contained. If the chassis is used for the return path, then the field can be rather extensive.

In a practical transmission line a small amount of power is dissipated in the conductors. The voltage drop in the conductors requires an E field in the conductors. This E field points in the direction of the conductors. Poynting's vector for this E field and the existing H field points into the conductor. At every point in space there is a small component of Poynting's vector that shows that a part of the power is being dissipated in the conductors.

RADIATION

TRANSMISSION LINE — POWER FLOW

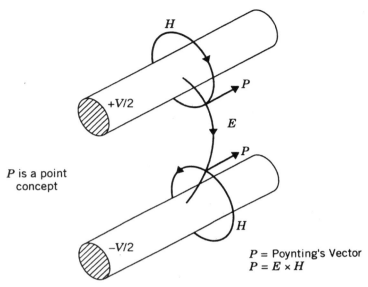

Figure 3.6 Poynting's vector and the transmission line.

3.10 RADIATION

The concepts of capacitance and inductance involve the storage of field energy. In many geometries the fields extends out into space. Within the context of circuit analysis the field energy is confined to the region of the component. In a sinusoidal analysis the tacit assumption is that this energy is stored and returned to the circuit twice per cycle. When the fields extend into space, this assumption may not be valid.

The energy stored in a field is proportional to the square of the field intensity. This means that the bulk of the field energy exists close to the source of voltage or current. When the field energy follows a slow sinusoidal pattern the energy flows into the field and back into the circuit with no difficulty. It takes time for a field to build up because the field can only propagate at the speed of light. The field that has propagated the farthest takes the longest time to return to the circuit.

A magnetic field that moves in space is associated with an E field. Both fields must be present to move energy. When an electric field moves in space an H field must be present. The fields produced from a sinusoidal source are the simplest to consider. In practice the fields of interference are pulse in character (ESD) or from rapidly changing voltages (digital signals). These signals are said to be rich in spectrum. Any analysis requires considering each part of the spectrum separately and then combining the results. The radiation from sinusoidal sources is obviously simpler to handle.

3.11 RADIATION FROM A DIPOLE

Solving the radiation problem involves solving Maxwell's equations. The simplest radiator to consider is a short conductor segment. This segment is a part of a larger antenna called a *dipole*. The radiation pattern for the entire antenna is found by summing the radiation from each dipole segment. The segment is ideal in that the conductor has no diameter and the rest of the current path is not considered.

The fields for a dipole element are given by

$$E = k_1 \left(\frac{\lambda}{2\pi r}\right)^3 + k_2 \left(\frac{\lambda}{2\pi r}\right)^2 + k_3 \left(\frac{\lambda}{2\pi r}\right) \tag{3}$$

$$H = k_5 \left(\frac{\lambda}{2\pi r}\right)^2 + k_6 \left(\frac{\lambda}{2\pi r}\right) \tag{4}$$

where λ is the wavelength in free space and r is the distance from the dipole. All vector notation and spherical coordinate notation has been left out to simplify the equations.

The term $\lambda/2\pi r$ repeats in every term. Near the dipole where r is small the first term in Eqs. (3) and (4) dominates. For large values of r the last term in Eqs. (3) and (4) dominates. When $\lambda/2\pi r = 1$, all of the terms contribute equally. The value for r is

$$r = \frac{\lambda}{2\pi} \tag{5}$$

This distance from the dipole is said to be the near-field/far-field interface. At distances greater than $\lambda/2\pi$ the E and H fields fall off linearly and the ratio of E to H is a constant. This ratio is called the *impedance of free space* or 377 Ω.

Inside the near-field/far-field interface the wave impedance rises. For small r the ratio of E to H is inversely proportional to r. At half the interface distance the wave impedance is doubled. The near field for a dipole is said to be a high-impedance field. It turns out that it is easy to shield against a high-impedance field.

3.12 RADIATION FROM A CURRENT LOOP

The radiating field pattern for a loop antenna is given by Eqs. (6) and (7).

$$E = k_1 \left(\frac{\lambda}{2\pi r}\right)^2 + k_2 \left(\frac{\lambda}{2\pi r}\right) \tag{6}$$

$$H = k_3 \left(\frac{\lambda}{2\pi r}\right)^3 + k_4 \left(\frac{\lambda}{2\pi r}\right)^2 + \left(\frac{\lambda}{2\pi r}\right) \tag{7}$$

These equations are similar to Eqs. (3) and (4) with the role of E and H interchanged. Note that the interface distance is still given by Eq. (5). The wave impedance for signals in the far field is still 377 Ω, and both fields fall off linearly with distance. In the near field the wave impedance is proportional to r. At half the interface distance the wave impedance is one-half. A near field is dominated by the H field and is said to be an induction field. This type of field is difficult to shield against.

For power-related phenomena the field from a loop is a near field. The magnetic fields from transformers and current carrying conductors are very difficult to attenuate. For these types of fields it is easier to bend the field away from an area than to attempt to shield or reflect the field from that area. A better approach is to configure the circuit geometry so that the magnetic field is not generated in the first place. The interface distance at 60 Hz is about 500 miles. This means that all power phenomena are near field in character. The subject of magnetic shielding is treated later.

The idea of a moving electromagnetic field having an impedance is a matter of convenience. If a measure is made of the E field, the corresponding H field can be calculated. The calculation must take into consideration any near field effects and if so the nature of the source. In the far field the multiplier is simply 377 Ω. Often in making measurements it is easier to convert one field to another to determine if a specification is met. The term *wave impedance* has no other physical significance.

Wave impedance has units of volts/meter divided by amperes/meter. The meters involved are actually at right angles to each other. Some prefer to say that the ratio is ohms per square rather than just ohms. This idea will make more sense when surface impedances are discussed later.

Radiation from loop areas occurs often in circuit design. Every signal and the return common is a current loop. The loops can be as small as the power and logic line on an IC or the loop formed by the leads on a component. In most cases, circuit loops are rectangular rather than circular.

In the interest of obtaining a worst-case value for radiation, the approach usually taken is to ignore polarization and direction. The assumption is made that all radiators contribute in an optimum manner. The data given in Figure 3.7 are for one radiator having a loop area of 1 cm^2. The nominal voltage is 1 V rms. If the actual voltage is 5 V, then the result is multiplied by five. Similarly, if the area of a loop is 3 cm^2, a factor of three is used. If 10 gates fire at one time, another factor of 10 is used.

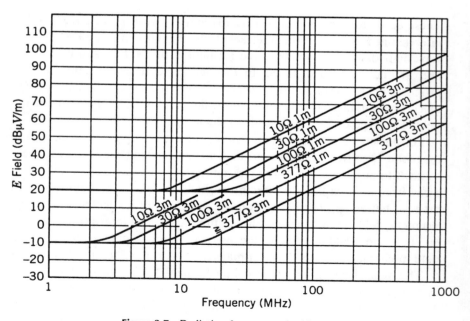

Figure 3.7 Radiation from a standard loop.

EFFECTIVE RADIATING POWER

The curves in Figure 3.7 are for two distances from the radiator. These distances are used in various standards that show permitted radiation levels. When considering a logic board it is fairly safe to assume that the field strength falls off linearly with distance. The complexity of dealing with a wide band signal and considering corrections for wave impedance is not worth the effort.

The frequency that should be used in Figure 3.7 is $1/\pi\tau_r$, where τ_r is the rise time. If 5-V logic is involved, the voltage multiplier is simply five. See Section 11.4 for a full discussion of frequency and amplitude selection.

The curves in Figure 3.7 are for different impedance levels. At low impedances the currents are higher and the radiation is greater. The impedance selected depends on the class of logic being considered. For CMOS logic the 377-Ω figure can be used. For TTL logic a figure of 100 Ω is acceptable. In practice the actual impedances are nonlinear and a selection is only an approximation.

3.13 EFFECTIVE RADIATING POWER

The radiated field pattern at a distance from an antenna depends on the directivity of the antenna. This directivity is usually described as antenna gain. When power is involved, directivity becomes an economic matter. For a radar signal, directivity makes it possible to get enough energy to the target to get a useful reflection. For a radio station it serves no useful purpose to broadcast into space or into a mountain.

When field energy impinges on a circuit the only thing that matters is the field strength at that point. If the assumption is made that the signal is broadcast in all directions, then the effective power of the source can be calculated. The actual power transmitted is reduced by the gain of the antenna.

The field strength at a distance from a transmitter can be easily calculated. The power crossing a sphere can be determined from Poynting's vector. Since the power is the same in all directions the power is

$$W = E \cdot H \cdot A \qquad (8)$$

where A is the area of the sphere. The area is $4\pi r^2$ and $E/H = 377$. Solving for E in terms of power W in watts we obtain

$$E = \frac{1}{r}\sqrt{30W} \qquad (9)$$

The power requirements for a radar signal is a good problem to consider. If the target field intensity is to be 1 V/m and the distance is 10 km, the effective radiated power required is

$$W = E^2 \cdot R^2/30 \tag{10}$$

or 2.88 MW. If the antenna gain is 100, then the power required during the pulse is 28.8 kW. If the power is pulsed 10% of the time, the average power required is 2.88 kW.

It should be remembered that the interference level from a radar beam results from a 2.88-MW source not from a 2.88-kW source. The effective radiated power from some military radars can exceed 1 GW. It is easy to see why electronic equipment on aircraft must be carefully designed. Aircraft often fly near transmitting antennas of all types.

4
FIELDS AND CONDUCTORS

4.1 INTRODUCTION

In electrostatics a fixed electric field results in charges on the conducting surface. When these same charges move in a conductor at dc, they flow inside the conductor. If the conductor is round and the current flows at dc, the current density is nearly the same for every element of cross section. Because all conductors have resistance, an E field must exist inside the conductor to cause the charges to move. The field is simply the volts per meter required by Ohm's law.

The flow of current in a conductor implies a magnetic field in that conductor. When the current changes, so does the magnetic field. This changing magnetic field in the conductor develops an E field (voltage) that opposes the flow of current. This has the effect of limiting the current that flows in the center of the conductor. This effect is known broadly as *skin effect*.

When a step voltage is applied to a circuit, the charges start to move from the surface to the interior. Progress is slowed by the resulting changing magnetic field. As time progresses, current flows in the entire cross section. When a sinusoidal current flow is impressed on a round conductor, the current establishes a fixed pattern where the current concentrates in the outer layers. The current distribution is described using Bessel functions. In most applications it is unnecessary to use this mathematics to get a good idea of how skin effect impacts the circuit.

Current patterns within conductors are complex and often ignored. For example, the radiating antenna problem did not address skin effect or resistance. In the reflection of waves from conducting surfaces the theory requires the conducting surfaces to be infinite in extent. This means that any attempt at measurement is a compromise. It is obvious that many compromises are necessary to get a picture of what is happening. This chapter is intended to give insight into an area that cannot be fully solved. The surfaces and geometries that are found in practice make it all but impossible to analyze with any accuracy. Fortunately, approximate answers are acceptable and this allows us to proceed.

4.2 OHMS PER SQUARE

The resistance of a cylindrical conductor is

$$R = \frac{\rho l}{A} \tag{1}$$

where ρ is the resistivity of the material, A is the cross-sectional area, and l is the length of the conductor. For copper, ρ is equal to 1.724×10^{-6} Ω-cm. When Eq. (1) is applied to a thin square of conductive material we obtain

$$R = \frac{\rho l}{A} = \frac{\rho}{t} \tag{2}$$

where t is the thickness of the material. The result is independent of the size of the square.

Equation (2) assumes that the current flows uniformly across the square and that the current uses the entire thickness. This point is illustrated in Figure 4.1.

The ohms per square for copper and steel is given in Table 4.1. An examination of the table illustrates how low this impedance is for very thin materials. A 1-mm-thick square of copper is 17 $\mu\Omega$ whether the square is a postage stamp or a full room. Again this impedance is only meaningful if current flows uniformly across the entire surface.

OHMS PER SQUARE

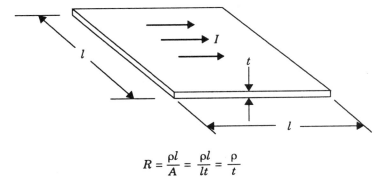

$$R = \frac{\rho l}{A} = \frac{\rho l}{lt} = \frac{\rho}{t}$$

R does not depend on dimension l.

Figure 4.1 The concept of ohms per square.

A thin sheet of copper (aluminum, steel) is an excellent conducting surface. These thin materials are not mechanically robust, and thicker materials are needed to be practical. In general a thicker material has a lower ohms-per-square value. Note that at higher frequencies the current cannot penetrate the surface and the ohms-per-square value rises. This topic is covered in Section 4.4.

TABLE 4.1
Ohms per Square for Copper and Steel

	Copper			Steel		
Frequency	$t =$ 0.1 mm	$t =$ 1 mm	$t =$ 10 mm	$t =$ 0.1 mm	$t =$ 1 mm	$t =$ 10 mm
10 Hz	172 µΩ	17.2 µΩ	17.2 µΩ	1.01 mΩ	101 µΩ	40.1 µΩ
100 Hz	172 µΩ	17.2 µΩ	3.35 µΩ	1.01 mΩ	128 µΩ	126 µΩ
1 kHz	172 µΩ	17.5 µΩ	11.6 µΩ	1.01 mΩ	403 µΩ	400 µΩ
10 kHz	172 µΩ	33.5 µΩ	36.9 µΩ	1.28 mΩ	1.26 mΩ	1.26 mΩ
100 kHz	175 µΩ	116 µΩ	116 µΩ	4.03 mΩ	4.00 mΩ	4.00 mΩ
1 MHz	335 µΩ	369 µΩ	369 µΩ	12.6 mΩ	12.6 mΩ	12.6 mΩ
10 MHz	1.16 mΩ	1.16 mΩ	1.16 mΩ	40.0 mΩ	40.0 mΩ	40.0 mΩ

4.3 REFLECTION

When a plane wave (a far-field wave) impinges on an infinite conducting surface, two things happen. Some of the wave reflects and some of the wave enters the conductor. The reflection process is similar to that of a transmission line terminated in a near short circuit. The energy is not dissipated but is instead turned back as a second wave. The boundary conditions require that the E field on the surface must be near zero. If there were a large E field on the surface, the surface currents would have to be immense. Poynting's vector for the reflected wave requires that the E field is reversed in sign while the H field is unchanged. An H field at the surface implies a surface current. Because the surface has a finite surface resistance, some E field must exist. It is this E field and the surface H field that propagates into the conductor. The wave that continues into the conductor is attenuated because of resistive losses. If the conducting surface were perfect, then wave energy could not penetrate the surface.

The reflected signal level is a function of wave impedance. A low wave impedance is a closer match to the surface impedance of the conductor, and less of the wave is reflected. This is why it is so difficult to shield against an induction or low-impedance field. Reflection attenuation is a measure of signal loss. It is the ratio of initial signal to signal continuing forward:

$$R_{\text{dB}} = 20 \log \frac{Z_W}{4Z_B} \tag{3}$$

where Z_W is the wave impedance of the impinging wave and Z_B is the surface impedance in ohms per square.

4.4 SKIN EFFECT

The wave that penetrates the surface is attenuated by losses in the material. The wave attenuates exponentially as it progresses. This means that the loss is proportional to the amount of signal present. Fortunately, absorption is not dependent on wave impedance. The attenuation factor is given by

$$A = e^{-\alpha h} \tag{4}$$

where h is the depth of penetration and α is a constant that depends on the material and the frequency of interest. The constant α is

SHIELDING EFFECTIVITY

$$\alpha = \sqrt{\pi f \mu \sigma} \tag{5}$$

where μ is the permeability, σ is the conductivity, and f is the frequency in hertz. The constants for copper are

$$\mu = 4\pi \times 10^{-7} \text{ H/m}$$
$$\rho = \frac{1}{\sigma} = 1.724 \ \mu\Omega\text{-cm}$$
$$\sigma = 0.580 \times 10^8 \text{ A/v·m}$$

The value α for copper is 0.117/mm at 60 Hz.

Skin depth involves the ratio of the square root of frequency. For example, the skin depth for copper at 600 kHz is 0.00117 mm. Equation (5) is approximately true for round conductor geometries. An accurate value requires the use of Bessel functions. In other geometries the skin depth can be even more difficult to calculate. The compromise approach that seems to be adequate is to use Eq. (5) for all conductor geometries.

In Eq. 5 the case where h equals $1/\alpha$ results in an attenuation of e^{-1}. This depth is called *one skin depth*. In decibels this an attenuation factor of 20 $\log_{10}e$, or 8.66 dB. In two skin depths the attenuation is 17.32 dB. Note that e is the base of natural logarithms, or 2.71828.

Skin depth plays a roll even at 60 Hz. In large transmission systems the conductors often have steel center cores to add strength to the conductors. Skin effect forces the current to the outside surface making the center less effective. The increased strength allows the support towers to be spaced further apart. This results in a lower transmission line cost per mile.

At 400 Hz, power cables are often made of several smaller parallel conductors. This makes better use of the available copper. Litz wire (multistranded conductors) are often used in air core transformers at high frequencies. The multistrand wire reduces the resistance of the wire at so that the coils can have a higher Q.

4.5 SHIELDING EFFECTIVITY

The shielding properties of a conductive material are described in terms of shielding effectivity (SE). A test is made using a square of material mounted in an opening in a large conducting bulkhead. A transmitter and receiver are set up on opposite sides of this opening. The signal attenuation that results when the test material is in place is the shielding effectivity of the material. Shielding effectivity is usually given in decibels. It is the total attenuation of an incident wave and includes refection(s) and absorption. SE takes the form of Eq. (6) when R and A are expressed in decibels:

$$SE_{\alpha B} = R_{dB} + A_{dB} \qquad (6)$$

R_{dB} is the loss due to reflection, and A_{dB} is the loss due to absorption. For very thin materials an added term may be needed to account for rereflections.

In practice the enclosures found in electronics are racks, sheets of steel or aluminum, metal boxes, plated or painted plastics, or conductive plastics. These enclosures are certainly not infinite surfaces. The shielding effectivity of these configurations is not easy to calculate or measure. It is common practice to assume the SE of an infinite sheet of material and accept the results. It is important that a margin for error be placed into the calculation to allow for errors introduced by this assumption.

Shielding effectivity is a function of permeability. For most magnetic materials the permeability falls off at high frequencies. For ferrites the conductivity is near zero and the material is not a reflective shield. Ferrites can serve to block current flow by transformer action when a conductor threads the core.

4.6 APERTURES

Electromagnetic fields can penetrate an enclosure through its apertures. The larger the aperture, the easier it is for fields to penetrate. Apertures are needed for seams, ventilation, controls, and monitoring components. A CRT monitor presents a large aperture as does a keypad.

The only safe way to treat apertures is to consider a worst-case scenario. The field is assumed to be polarized (the direction of the E field) for maximum coupling. The angle of arrival is assumed to be worst case or head on. If the polarization is off by 45°, the result would be a 3-dB error. For a worst-case analysis this error is not significant. The same thing can be said about the angle of arrival. If the interference is low enough after making worst-case assumptions, the problem is solved.

A very thin aperture such as a seam probably provides additional attenuation. In a worst-case approach the gap dimension of a thin seam has no bearing on the matter. The length of the seam is the dimension of interest.

The penetration of the field is proportional to the half wavelength of the aperture. At 10 MHz the half wavelength is 15 m. A seam that is 15 cm long attenuates the field by a factor of 100. An E field that is 10 V/m outside the enclosure is 0.1 V/m inside the enclosure.

The fields inside an enclosure are influenced by rereflections by component locations and by various losses. A field penetrating an aperture is not a plane wave. All of these effects are ignored in the spirit of a worst-case analysis.

4.7 INDEPENDENT AND DEPENDENT APERTURES

Apertures that are surrounded by a metal surface are called *independent apertures*. If two widely spaced apertures are involved, the fields simply add together. If one opening allows 0.1 V/m to enter and a second opening allows 0.2 V/m to enter, the assumption is made that the total field strength is 0.3 V/m. Field levels expressed in decibels must be converted to absolute values before they can be added. When there is doubt about the surface area around an aperture, it is wise to assume that the aperture is independent.

When apertures are arranged so that there is no free flow of current around each port, the effect is that of one aperture. These openings are called *dependent apertures*. Two pieces of metal held together by many screws (a seam) form a single aperture. The current cannot flow independently around the screws at the point of contact. A tight group of ventilation holes forms one dependent aperture. A honeycomb of conductors forms independent apertures because current can circulate around each tube of the honeycomb.

A grid of wires is a dependent aperture. The field penetration involves the dimensions of one square. The wires making up the grid must be bonded at every crossing. Aluminum screening is unsatisfactory because of oxidation. Chicken wire is acceptable because wires at each intersection are soldered together.

4.8 CLOSING APERTURES

Seams are best closed by a conductive gasket. A gasket must make good contact around any involved perimeter. This means that the surface cannot be anodized or painted. The surface should be plated to avoid oxidation. The gasket must be under uniform pressure to guarantee a contact along its length. Gaskets are often deformed when installed. If an opening needs servicing, an appropriate gasket should be used or the gasket should be replaced.

Doors form an aperture that can require special treatment. Fingerstock is frequently used, although a gasket under pressure can be adequate. Fingerstock has the problem of catching on clothing. If the fingerstock is mounted inside a U channel, this problem is resolved.

A grid of wires provides an excellent shield. If the grid is not bonded around its perimeter, an aperture results and the qualities of the grid are lost. If a grid has a wire spacing of 1 mm and the bond around the perimeter has a gap of 1 cm, the aperture is that of the gap and not that of the grid.

The opening for a CRT monitor is large. Aperture closure using a wire grid is not satisfactory because of the Moiré patterns that form. Conductive glass provides a limited closure. When the conductive surface is adequate, very little light can penetrate. One solution involves limiting the aperture to the diameter of the neck. This does not control the radiation that is generated by the beam itself. If a surface-mounted monitor can be provided, it avoids all of these issues.

4.9 WAVEGUIDES BEYOND CUTOFF

Any aperture with depth forms a waveguide. A waveguide provides effective attenuation at frequencies below cutoff. The lowest frequency that can propagate in a waveguide involves the largest dimension of the opening. At frequencies below one-half wavelength the waveguide attenuates a field exponentially.

The attenuation in a waveguide in decibels is given by

$$A = 30\ h/l \tag{7}$$

where h is the length of the waveguide and l is the aperture opening. This attenuation is independent of frequency, assuming that the waveguide is operating below cutoff.

An example can illustrate the attenuation of a waveguide. An round aperture has a diameter of 1.5 cm. The half wavelength at 10 MHz is 15 m. Assume the external field strength is 10 V/m. The field at the entrance to the aperture is attenuated by a factor of 1000 (60 dB) or 0.01 V/m. Without a waveguide this is the field strength inside the enclosure. In this example the field is further attenuated as it travels down the waveguide. If the waveguide is 3 cm long, the additional attenuation using Eq. (8) is 60 dB. The field inside the enclosure is thus attenuated by a total of 120 dB or by a factor of 10^6. To achieve this same attenuation without a waveguide the hole would have to be 0.0015 cm in diameter. It is easy to see that adding depth to an opening can be very effective. In designing hardware, the depth of a flange can often make all of the difference.

4.10 A REVIEW OF FIELDS ENTERING AN ENCLOSURE

A worst-case analysis is the only practical way to approach the field levels that enter an enclosure. Fields can enter by penetrating the conductive walls. This entry depends entirely on skin depth. This field is added to any fields that penetrate through apertures. The apertures can be of two types, dependent or independent. The closure of apertures can be handled by gaskets, wire screens, or waveguides beyond cutoff. The importance of closing an opening totally cannot be overemphasized.

Any conductor that enters the enclosure can bring field into the enclosure. This completely nullifies the effect of closing apertures or providing waveguides beyond cutoff. All conductors must be filtered to the outside of the enclosure to avoid contaminating the enclosure. This topic is discussed in more detail when power-line filters are considered. A conductor threading a waveguide entry nullifies the wavequide. It then becomes a piece of coaxial cable, and fields can enter at all frequencies. A conductor that enters an enclosure and is then bonded to the inside of the enclosure will radiate energy into the enclosure. The loop that is formed is the radiator.

4.11 THE COUPLING OF FIELDS TO CIRCUITS

When an electromagnetic field enters an enclosure, the wave energy couples to the conductor geometry because this is an easy path to follow. Inside of most electronic devices the coupling is proportional to loop area. The loops can involve wiring in circuits as well as conductors routed over the conductive enclosure. Wiring includes shields, commons, power leads, and signal leads.

The simplest way to approach the coupling mechanism is to consider a long pair of conductors separated by a distance h. The field of interest propagates along the path of the conductor pair as in Figure 4.2. The field induces a voltage into the loop that appears in series with any other voltage sources. This voltage can be determined from Faraday's law using the changing H field. The result is proportional to the loop area of the circuit. The same answer can be obtained by considering the E field alone.

At any instant in time the E field at one end of the conductor pair is different than the value at the other end. Assume that the E field polarization is in the direction of h. The sum of voltage around any loop must be zero. The E field produces no voltage along the length of the run. At any moment in time the voltage induced in the loop is the difference in E field at the ends of the run times the height h. For example, if the E field is 10 V/m at one end and -5 V/m at the other end and $h = 0.01$ m, the induced voltage is 0.15 V.

A maximum signal couples to the conductors when the conductor length is one-half wavelength. The coupled voltage is twice the peak E field times the distance h. For lengths less than one-half wavelength the voltage is proportionately lower. At 10 MHz the maximum coupling occurs for a length of 15 m. At 15 cm the coupling is simply 0.01 times the peak value.

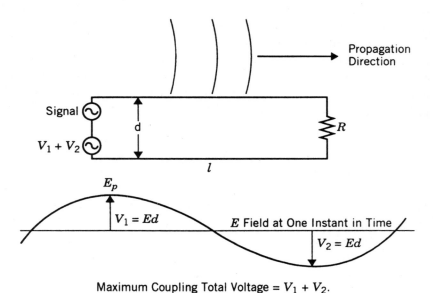

Maximum Coupling Total Voltage = $V_1 + V_2$.
When l is a half wavelength.

Figure 4.2 Field coupling to a conductor pair.

THE FIELDS IN A ROOM

If the field strength is 10 V/m and $h = 0.01$ m, the coupling is 2 mV. The coupled voltage is proportional to the height and length and thus proportional to the loop area.

The method shown above is a worst-case analysis. It is an easy calculation to make. It assumes an optimum propagation path and an optimum wave polarization. If the length of the conductor pair is greater than one-half wavelength, theory says the coupling is reduced and is zero at one wavelength. This assumption of cancellation is not in the spirit of a worst-case analysis. The safest assumption to make is that the coupled voltage is maximum at one-half wavelength and stays at this maximum value for cable lengths greater than one-half wavelength.

The coupling of fields to conductors over a ground plane is proportional the conductor length and the height above the ground plane. The coupling is again proportional to area. When the conductors are one-half wavelength long the coupling is again optimum. Above this length the worst-case analysis approach requires that this value of coupling be used. Coupling to a group of conductors is called *common-mode coupling* (see Section 4.13).

On printed circuit boards the loop areas formed by traces over the ground planes are very small. Coupling problems are more apt to occur when ribbon cables leave a board. Here the loop areas formed with a ground plane can be significant. Fortunately, digital signals are at a relatively high level, and small amounts of coupling do not interfere. In the case of the earth it is often difficult to determine a proper height. A figure often used is one skin depth at the frequency of interest. This will vary depending on moisture content.

4.12 THE FIELDS IN A ROOM

A typical room in an area of electrical activity is bombarded by radiating sources. These sources include radio, television, police radio, and any local radar. Unless a screen room is available, there is no simple way to avoid these fields.

Interfering fields can be carried by the conductors in a power system. These fields are added to the fields used for power transfer. Fields carried by these power conductors are contained when the leads are carried in conduit. This containment stops at power transformers located in various pieces of equipment. Power currents that do not flow in conduit can release fields that couple to nearby circuitry.

The neutral conductors in power distribution (external to facilities) are multiply earthed at each service entrance. Unbalances in three-phase loading cause neutral currents to flow in the earth. The low-impedance paths are obviously earthed conductors such as building steel, gas lines, and water pipes. These conductors enter each facility, and the fields associated with this current flow are a source of interference.

In each transformer there are capacitances from primary to secondary coils. Current flowing in this capacitance can follow paths provided by signal cables. The return path is obviously complex, and current flows in large loop areas; thus there are fields in the room. This current can be somewhat controlled by line filters and transformer shields. This topic is covered in later chapters.

Cables that leave a facility (input or output cables) can bring field energy into a room. The energy can travel between conductors and between the conductors and the ground plane. This energy cannot be rejected except perhaps at the walls of a screen room. It is obvious that a conductive enclosure for electronics can constitute a miniature screen room.

The fields in a room are not shorted out by building steel or by conduit. It follows that large conductors connecting specific points in the room will not eliminate these fields. It also follows that a grounding conductor to earth will not attenuate these fields. The approach that must be taken in the design of equipment is to accept the presence of fields and design to reject their influence. The adage that says "If you can't beat them, join them" has meaning. There are ways to distribute power to equipment to reduce these problems. This is discussed later.

It comes as a surprise to many engineers that there are voltages between the metal covers of grounded pieces of test equipment on a laboratory bench. This voltage can be measured with a voltmeter or seen on an oscilloscope and may be several volts. This voltage is just a measure of the fields that exist in the loop area formed by the measuring device, the equipment, and the power connection. There need not be current flowing in this loop for there to be a potential difference. Some of this difference can be the result of neutral current flow.

The loop area formed by voltmeter leads or by any signal path can couple to field, and this results in a voltage. Measuring the voltage between various conductors can lead to incorrect answers if the loop area formed by the voltmeter leads is not considered. If a voltmeter touches two points on a ground plane, the answer will depend on the field strength in the loop that is formed. To avoid this problem, always route the leads for a minimum loop area.

COMMON MODE AND NORMAL MODE

The electric field pattern in a room is modified by the presence of every conductor, and this includes the human body. A weak TV signal is modified as a person moves about a room. The movement of a hand into a piece of hardware will affect signal coupling. At low frequencies the body can be looked at as a walking capacitor. The body has a capacitance of about 250 pF to the floor. Fields that terminate on the body cause surface currents to flow. This flow is an interfering signal when body functions are monitored electrically.

The surface resistance between points on the body is a function of skin condition. If the skin is dry, the resistance between two contacts of one-fourth square inch can be 100,000 Ω. When the contacts are moist the resistance can drop to 2000 Ω. Measurements made subcutaneously are in the order of 100 Ω. This implies that the body is a fairly good conductor and does indeed reflect most low-frequency E fields in the room.

4.13 COMMON MODE AND NORMAL MODE

The signal of interest is often called the *normal-mode signal*. Other names include differential signal, difference signal, or transverse signal. This desired signal is often imbedded in an average or undesired signal. This average or undesired signal is often called a *common-mode signal*. In telephony this unwanted signal is called a *longitudinal signal*.

One way to observe a common-mode signal is to note the signals present when the normal signal is zero. There will often be a response to these unwanted signals where ideally there should be no response. The rejection of these common-mode signals is an important performance specification.

The common-mode signal present in many applications is called a *ground potential difference*. When a signal is sensed on one structure and observed relative to the ground on a second structure, the voltage between the two grounds is called a *common-mode voltage*. This voltage is a measure of the field in the loop formed by the signal cable and the conductive path between the two structures.

A common-mode voltage can be defined as the average voltage with respect to a reference conductor where the signal of interest is set to zero. There can be several common-mode signals in effect at the same time. For example, the utility power represents a common-mode signal with respect to a signal common. In an amplifier the dc power supply voltages are a common-mode signal. In a strain-gage amplifier the gage excitation is a common-mode signal. In each of these examples the gain to these common-mode signals should ideally be zero.

If the definition of common-mode voltage is applied to a single-phase power entrance, the average voltage is one-half the line voltage. This voltage is certainly necessary if there is to be a source of power. Because of the way a power transformer is built, the turns of wire on the primary coil bury the midpoint inside the coil. This means that the end-turns of the coil are either associated with the power ground or the high side of the power line. The capacitance between coils in a transformer is in series with an effective voltage that depends on coil geometry. In this example the average voltage is not involved in any coupling.

When a pair of conductors carry a signal over a ground plane, the involved loop area can couple to any electromagnetic field. This field couples to both signal conductors in the same sense. This signal is a common-mode signal. For analog applications this signal is usually out of band. This signal is brought into the equipment on both the input leads and on any input shield conductor. Equipment designed to function with long input lines should consider the impact of this type of common-mode coupling. When a shield carrying high-frequency field energy enters a circuit, the effect can be to overload various stages. The input stages are of course the most vulnerable.

The neutral voltage drop in a large facility is an average voltage applied to the primary coil of each transformer in every device. This is a common-mode signal. The neutral voltage drop is often noisy and rich in power harmonics. This noise voltage causes reactive current to flow in secondary commons, and this adds noise to any secondary signals. Any regrounding of the power system is illegal and unsafe because it impacts the fault protection system. A separately derived power system can be used to solve this type of common-mode problem.

5
ELECTROSTATIC SHIELDING—I

5.1 INTRODUCTION

Interference at low frequencies can often be handled without considering the magnetic field. In this chapter, cases are considered where interference can be blocked by interposing a conductive barrier. As simple as this idea is, there are many mechanisms that must be considered. The basic idea involves placing the circuitry into a metal enclosure (box). External electric fields must terminate on the outside surface and therefore cannot couple to the inner circuitry. The "box" can take on any shape, including a cylinder (shielded cable). The "box" also limits any internal fields from leaving, and thus there is no coupling to any nearby circuit (see Section 1.24).

The magnetic field must be considered when there are very high currents, when circuits are near power transformers, or if input circuitry is near a ventilation fan. There are a few examples where the low-frequency magnetic fields dominate. Examples include nearby VLF transmitters (10 kHz), magnetic resonance scanners and magnetic levitation such as for trains. The coupling to these magnetic fields involves Faraday's law (see Section 2.5).

The ability to shield effectively is related to wave impedance. At frequencies below 100 kHz the wave impedance is usually very high and the majority of the field energy is simply reflected. The near-field/far-field interface for power and its harmonics is at a distance greater than 500 miles. This means that for voltage (E fields) the wave impedance is high and any metal plate will serve as a reflector. For power currents the field is a near induction field and the wave impedance is low. Shielding against this type of field is difficult.

5.2 THE CIRCUIT IN A "BOX"

When a circuit operates in an electrostatic enclosure the voltages between conductors establishes a very complex E field pattern. If there are voltages, there are fields and some of this field terminates on the inside surface of the "box." The energy stored in this field pattern can be attributed to energy stored in various self- and mutual capacitances. It is difficult to analyze the way the circuit behaves by looking at field lines. The behavior of the circuit becomes clearer when we draw a circuit diagram. Figure 5.1 shows the field lines, their capacitive relationship, and their equivalent circuit diagram. Note that there is capacitive feedback around the circuit, which is usually not desired. The triangular symbol in the box could be any circuit, amplifier, or active device.

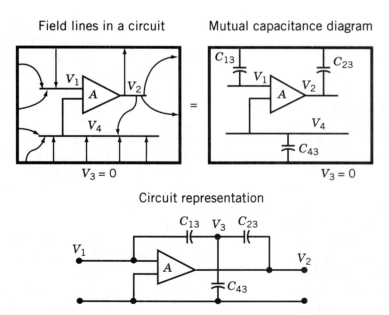

Figure 5.1 Going from field to circuit.

THE CIRCUIT IN A "BOX"

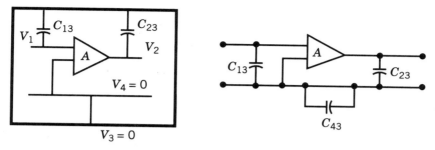

Figure 5.2 Grounding the circuit.

To eliminate the feedback circuit it is necessary to place a connection between the circuit and the "box." This is known by any electronic designer as "grounding the circuit." This connection is shown in Figure 5.2.

If a single lead is brought outside the box (not connected), then the capacitances to external conductors adds to the difficulty. The parasitics are shown in Figure 5.3. When the circuit equivalent is drawn, it is apparent

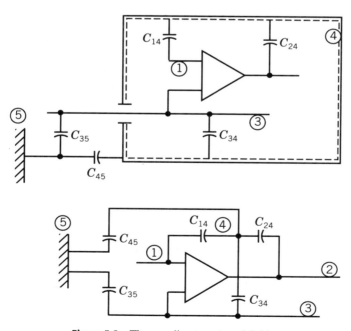

Figure 5.3 The coupling to external fields.

that the fields inside and outside the box are now interactive. When there are changes in the external field, there are interference signals at the input to the amplifier. In a crude way the external lead is said to be an antenna. The preferred explanation involves a simple capacitive divider structure.

When the circuit is "grounded" as in Figure 5.2 the undesired feedback is removed and external fields do not couple into the circuit. This grounding connection is often essential to the correct operation of a circuit. If all circuit conductors stay totally inside a controlled electrostatic region, this grounding connection may not be required.

5.3 EXTERNAL GROUNDING OF ONE CONDUCTOR

The next level of complication occurs when the common conductor for the circuit in the box leaves the box and is "grounded." "Grounded" here means a connection to some external conductor. This external conductor might be earthed, but this is not necessary to illustrate the new problems. The new connection modifies the field pattern between the box and the external "ground." With this connection the potential difference between the "box" and the external conductor is zero.

It is easy to see that this grounding arrangement is a source of interference. Figure 5.4 shows the circulation of current that results from the capacitances and external potential differences. This capacitive divider circuitry provides a good picture of phenomena below 100 kHz.

If the "grounded" conductor leaving the "box" is a signal conductor, then any current flow in this conductor will develop a potential drop. This potential drop is added to any signal. On a long conductor it is easy to have a resistance of several ohms and an inductance of over one millihenry. If a current of 1 mA flows, a voltage drop of several millivolts results. This voltage is interference, and it obviously depends on frequency. For large signals and/or for short runs the level of interference may not be objectionable.

5.4 THE GAME

There is a game that must be played. Circuits must be structured so that parasitic currents do not flow in signal conductors. This is not guaranteed by shielding alone. Where a shield is connected is often more important than whether it is connected. This type of information is not always provided by a circuit diagram. The heart of interference control is circuit geometry.

THE GAME

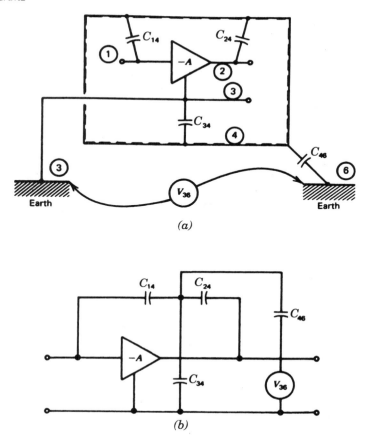

Figure 5.4 (a) Grounding one conductor. (b) Circuit equivalent of Figure 5.4a.

As the frequency of interest rises, geometry becomes more and more important.

In the circuit of Figure 5.4 a mistake was made. The circuit common (reference conductor) was connected to the shield (box) inside the box. If the connection is made at the point where the circuit common "grounds," then the problem is solved. Figure 5.5 shows a proper grounding. Note that the external field now circulates current in the shield not in the signal conductor. This is a good low-frequency solution to grounding a signal common. The rule is simple:

Connect the shield to the circuit common where the signal lead contacts ground.

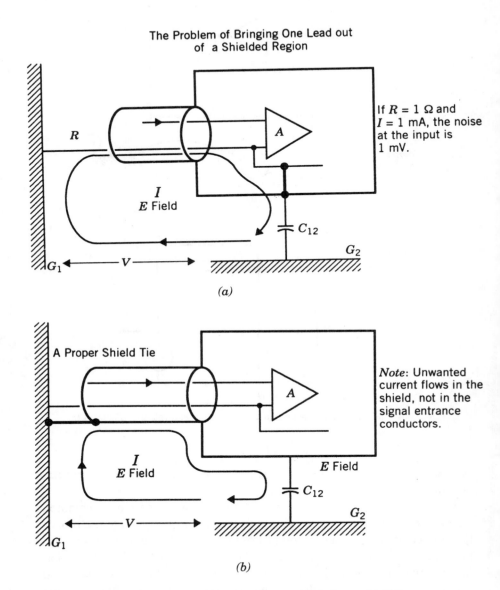

Figure 5.5 (a) An improper shield tie. (b) A proper shield tie.

5.5 THE TRANSFORMER CONNECTION

Battery operation of equipment is usually not practical. Equipment used in facilities for measurement, signal processing, signal analysis, and recording use power transformers. The presence of a transformer and its associated high voltages can easily contaminate the "box." The transformer primary is associated with a ground connection (earth) at the power service entrance. The secondary coils are connected to the circuit common after the power is rectified, filtered, and regulated. The transformer allows interference currents to flow in a grounded input signal conductor. The circulation path is shown in Figure 5.6.

The capacitance between coils in a transformer can range from 100 pF to 1000 pF. At 60 Hz, 1000 pF has a reactance of 2.65 MΩ. If the primary voltage is 120 V, the resulting current flow can be as high as 45.24 μA. If the input conductor has a resistance of 2 Ω, the resulting voltage is 90.5 μV. In many applications this is not a problem. If the signal of interest is

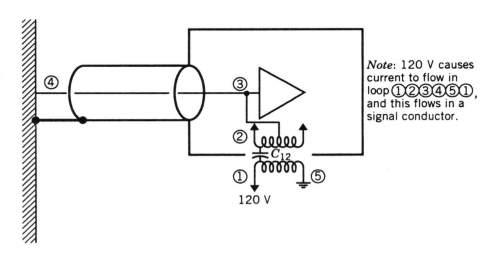

Figure 5.6 The path for interference current flow when a transformer is added to a box.

10 mV, the peak interference signal is 0.127 mV or 1.2%. This level of interference could thus be a problem.

There are several more voltage sources in series with the transformer coil-to-coil capacitance. The largest voltage is the primary voltage itself. This voltage may be less than the full primary voltage if the turns adjacent to the secondary are associated with the low side (grounded side) of the power source. Voltages across the primary coil can include any differential noise caused by load demands in other equipment. There are at least two common-mode sources in series with the transformer capacitance. The first is any common-mode signal using the power conductors as a transmission line. The second is the voltage drop in a neutral or low side power conductor (grounded conductor). This voltage is the result of load currents flowing in the line reactance. The smaller voltages in series with the primary voltages are apt to be higher in frequency and can cause higher currents to flow in the transformer capacitances. These sources include harmonic voltage content, transients caused by step load changes, and fields being transported by the power conductors as a transmission line.

With these sources of interference always present, techniques are needed to keep this current flow out of signal conductors. Some of these circuit techniques are discussed in later chapters. It is a valid exercise to see what can be done to shield the transformer in Figure 5.6 to limit or control this unwanted current flow. This exercise illustrates how transformer shields work. Also there are a few designs where this type of shielding is necessary. These transformer shields are often called Faraday shields to indicate that they control the electric field. They are simply wraps of thin conducting foil between coils of the transformer. These wraps must be insulated to avoid forming a shorted turn. A lead soldered or in contact with this foil is brought out for external connection. A boxed shield refers to wrapping the foil around the ends of the coil. For maximum shield effectivity, the leads to the coil must also be shielded. The shielding material is often copper, but it can be aluminum.

5.6 THE SINGLE TRANSFORMER SHIELD

The temptation is to add an electrostatic shield between the primary and secondary coils and use this shield to provide "closure" for the box. This solution is shown in Figure 5.7. There are several problems associated with this solution. There is a leakage capacitance from the primary coils to the secondary coils. This is shown as C_{36}. For a simple shield this capacitance

THE SINGLE TRANSFORMER SHIELD

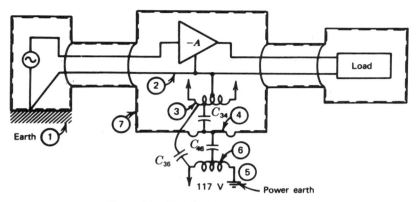

Figure 5.7 The single transformer shield.

might be as low as 5 pF. The current in the input conductor is reduced but certainly not to zero. The secondary voltages now circulate current in the loop ① ② ③ ④ ⑦ ①. The capacitance C_{46} can be several hundreds of picofarads.

The most severe problem occurs if there is a fault from the primary turns to the transformer shield. The fault path would involve the input shield and any input grounding. If the input circuit is not earthed, then there is a safety hazard involved. If this external ground were earthed, the loop that results is inductive and this limits the fault current. Again a safety hazard could result.

If one shield is available in a transformer, it should connect to the equipment grounding conductor (green wire/safety wire). If the chassis or box is also connected to equipment ground, then obviously the shield can connect to the box. The correct shield connection is shown in Figure 5.8.

When all of the connection possibilities for a single shield are considered (aside from the safety issue), there is no "good" solution to the problem. For every connection, the transformer voltages can still circulate current in the input common conductor. It actually takes three shields to solve the general problem, and this is an expensive solution. Other circuit techniques are available and are discussed later.

The sensitivity to current flow in an input conductor depends on the line length, the frequencies of interest, the signal level, and the accuracies desired. The currents just discussed will obviously flow in an output conductor. Generally, output leads are shorter and carry full-scale signals. For this reason, many of the issues raised regarding interference caused by unwanted

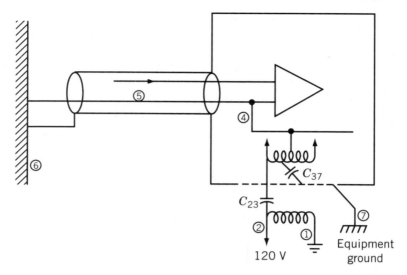

Figure 5.8 A proper transformer shield tie (electrically safe). Note: 120 V causes current flows in path 1 2 3 4 5 6 7 1. Current flows in path 7 3 4 5 6 7 through ground potential difference V_{67} and secondary voltage.

current flow are of little concern. A 1-mV signal is a small error for a 10-V full-scale signal. In some A/D designs the objective is to provide resolution greater than 16 bits. This is one part in 65,000. For a 10-V signal this is 0.15 mV and 1 mV of noise is impacting.

5.7 THE THREE-SHIELD SOLUTION

The three-shield solution to the power transformer entry is not often required. When three shields are required, it is necessary to understand how they function. The interference currents that flow in coil-to-shield capacitances do not enter the signal conductors. The interference current that flows in the primary leakage capacitance does not flow in a signal conductor. The shield connections are shown in Figure 5.9. The direct coil-to-shield capacitances are C_{34} and C_{27}. The leakage capacitances are C_{26} and C_{36}. The current in the direct capacitance path C_{27} flows in circuit loop ① ② ⑦ ①. The current in the direct capacitance path C_{34} flows in circuit loop ③ ④ ③. The current in leakage capacitance C_{26} flows in circuit loop ① ② ⑥ ①. The current in leakage capacitance C_{36} flows in circuit loop ③ ⑥ ⑤ ④ ③. Current here flows in circuit conductor ⑥ ④. Fortu-

THE SINGLE-ENDED INSTRUMENT

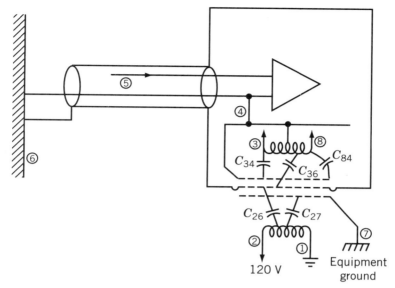

Figure 5.9 The three-shield power transformer solution.

nately the voltages on the secondary coil are smaller than the primary voltage. Often this secondary current can be nulled out by adding a trimmer capacitor C_{84}. If the transformer is built so that the centertap is taken from the outer coil layer, the secondary voltage in series with the leakage capacitance can be further reduced. In some applications the secondary coil may require a boxed shield to further limit the leakage current.

Multishielded transformers are expensive and are avoided unless there are no alternatives. The term "isolation" is often applied to these transformers. In the general sense these transformers do isolate the device from power-line interference. The isolation and shielding discussed here can only apply to one circuit. One transformer cannot be associated with two devices because this would require the circuit commons to be tied together. Isolation transformers for a computer installation or for an electronic facility involve a different approach to shield connections. This is discussed in Section 9.9.

5.8 THE SINGLE-ENDED INSTRUMENT

The grounded input circuit was discussed in Section 5.3. The input common lead was brought out from the box and "grounded." Here "grounding" means a connection to some device, earth, or structure. This was a very simple circuit and did not consider the fact that the output of the device may also require "grounding."

If the output were terminated in a light galvanometer, then output grounding would not be necessary and the input circuit could be grounded. If the output were a loudspeaker, then output grounding may not be necessary and again the input circuit could be grounded. For the general instrument both input and output signal lines may require grounding. Grounding of both input and output circuits causes the classic "ground loop," and this can result in a very noisy signal. This problem and its implications are topics discussed in the next sections and in the next chapter.

Digital interconnections form the classic single-ended circuitry. Logic input and the output circuits are both referenced to a single power supply common. Each signal is related to a source ground, and the output signals are referenced to the ground at the next circuit. For short distances the ground loops that are formed are acceptable. Fortunately the signal levels are volts, not millivolts, and interference is usually not an issue. This problem becomes more severe when long buses are used or when there is communication between separately powered devices. Long signal runs and any resulting current loop can also result in radiation.

5.9 SINGLE-ENDED SIGNAL SOURCES

The term *single-ended* refers to a signal present between two conductors. Often the signal source has a low side that may be connected to the structure being monitored or it may be connected to a shield. Examples might be a piezoelectric transducer, a magnetic pickup, a thermocouple, or a microphone. The shielding and grounding connections may be totally defined when the device is mounted. There are devices where the user makes some or all of the connections.

Single-ended signals can arise from a circuit or an instrument. Usually these signals are high level and have a low source impedance. One side of the signal is connected to the source ground or common. Shielding may not be required for these signals. Examples might be the output of a preamplifier or a digital logic line. Within a complex instrument there can be combinations of single-ended and balanced signals. Internal circuit shielding is rarely required.

A single-ended signal source is often referred to as an unbalanced source. If the signal arises from a grounded source, the impedance to that ground from the two signal leads is different. The impedance to ground can be active as in the case of an amplifier output lead.

The power line is a source of unbalanced power. One side of the power line is earthed at the service entrance. The impedances to earth from the two power leads are not equal.

5.10 BALANCED SIGNAL SOURCES

In audio work, signals are often transported over a balanced transmission line. The balancing of signals implies two oppositely moving voltages with respect to a reference conductor. This reference conductor could be the centertap of a transformer. The impedance to the reference conductor from the two signal leads should be equal. The reference conductor is often grounded at the source of signal.

In telephony, voice signals are transmitted in balanced form over open wire. A single-ended signal can be balanced by putting the signal through a transformer with a centertapped secondary. Signals can also be balanced by using an active circuit. Balancing implies equal source impedances and signal levels with respect to a reference conductor. A balanced line allows the rejection of common-mode signals. In a balanced system, common-mode coupling affects each side of the line equally and the voltage difference is not changed.

In instrumentation, many signals are generated in a Wheatstone bridge configuration. A Wheatstone bridge consists of four impedances. The bridge is excited by a voltage V (ac or dc) as in Figure 5.10. The signal is observed across the two nonpowered corners of the bridge. Signals are generated when elements of the bridge change value as the result of some external

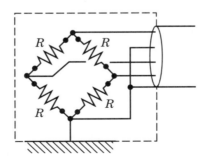

Figure 5.10 A Wheatstone bridge.

stimulus. If the bridge is used to measure strain, then the gage elements are resistances that change as a result of a dimensional change on some structure. The resulting signal is balanced with respect to the excitation signal.

Balanced resistive structures can be used to measure stress, strain, twist, position, temperature, pressure, and vibration. The signals are usually small (millivolts), and the bridge requires dc excitation. One of the excitation leads is usually grounded. The signal is balanced with respect to this ground. When the bridge resistances change, the balance is slightly affected. This can be a source of interference. The processing of these balanced signals to avoid interference is treated in later chapters. In some applications the frequency band of interest can extend from dc to 100 kHz. The connections of shields for various transducer configurations are discussed later. The shielding of the excitation power supply is discussed in Section 7.4.

5.11 THE TWO-GROUND PROBLEM

In general, signals must be sensed in one electrostatic region, amplified, conditioned, and then recorded or observed in a second electrostatic region. The electronics must, in effect, bridge the fields that exist between sensing and recording devices. The rules generated so far require that circuits be put in a "box" and that only one lead can be brought out to be grounded. With two grounds there must be two "boxes." The question that remains is how to communicate between "boxes." The input circuits must include all signal conditioning and excitation. The output section must include at least the output stages. The gain could be distributed between the two boxes. This general arrangement is shown in Figure 5.11. The power supplied to the input circuit requires a transformer with three shields. This is particularly true if small unbalanced signals are to be amplified. The output power is less critical because interference current in the output common lead does not generate a significant voltage drop. In the output circuits of low-level instruments, it is desirable to limit the maximum contributor which is the primary to secondary capacitance. A single shield connected to the safety ground limits this current to acceptable limits.

The excitation power supply within the input "box" requires special consideration, and this will be discussed in Section 6.7. Putting this problem aside for the moment, the most economical solution to the two-box problem is the eliminate active circuits in the input box. This eliminates the need for a three-shield transformer. The box itself cannot be eliminated because

THE TWO-GROUND PROBLEM

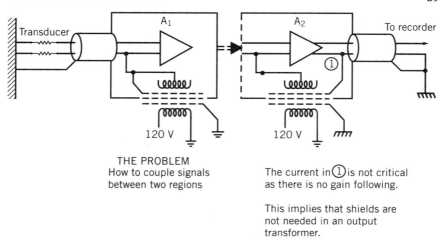

Figure 5.11 The two-box arrangement.

the transducer, any signal conditioning, the excitation regulator, and the input leads must be inside the box. This solution is shown in Figure 5.12. Note that the input box is the input shield. It is connected to the signal where the signal grounds. This shield is commonly referred to as a "guard shield." In effect, it is used to guard the input signal with the zero of

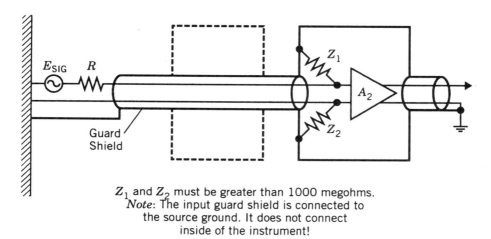

Z_1 and Z_2 must be greater than 1000 megohms.
Note: The input guard shield is connected to the source ground. It does not connect inside of the instrument!

Figure 5.12 The practical two-box problem.

signal potential. Later, when strain gage (Wheatstone bridge) signals are considered, the input cable can have up to 10 conductors.

5.12 THE FUNDAMENTAL INSTRUMENTATION PROBLEM

In Figure 5.12 the input circuit for box 2 has input impedances Z_1 and Z_2. It has become standard practice to specify this circuit configuration with an unbalance resistance R of 1000 Ω. In previous discussions, interfering input current would flow in a grounded input conductor. The impedance might be a few ohms. Now consider the effect if the interfering current flows in the unbalance resistance. The effect is amplified by 1000.

If the gain is 1000 and the interference is to be held below 10 mV in the output, the maximum input interference is 10 μV. This is a current of less than 10 nA in 1000 Ω. If the common-mode interference signal (ground potential difference) is 10 V, the parallel input impedance of Z_3 and Z_4 must be greater than 1000 MΩ. Note that this input impedance is required to meet a common-mode rejection specification not to allow operation from high source impedances.

One thousand megohms at 60 Hz is the reactance of 2 pF. This means that the mutual capacitance out of the input guard shield to local grounds must be held to below 2 pF. This further means that the leads inside the instrument must be well-guarded. Inside the instrument, input shielded cables are unnecessary if careful layout techniques are used. Active input impedances greater than 1000 MΩ are easily attainable. This means that it is practical to solve the two-box problem without active circuits in the input box.

Basic to this approach is the issue of where to connect the guard shield. If the circuits are balanced, then the common-mode voltage is not converted to normal signal and the grounding rule can be violated. Unfortunately the very signal itself provides an unbalance to the bridge, and this adds error to the signal. Obviously for unbalanced signal sources the guard shield must be properly connected or significant interference will result. The noise that results can be filtered out, but this is a dangerous approach. If the noise overloads the instrumentation prior to filtering, the data will be corrupted.

The fundamental instrumentation problem is how to sense signals in one electrostatic region, amplify and condition the signal, and observe or record the signal in a second electrostatic environment. There are many factors to be considered: the signal level, the source impedance, the balanced or unbalanced aspects of the signal, the bandwidth of interest, the common-mode levels that might be encountered, and the accuracies and stabilities desired. All of these factors must be considered in the light of costs. When many signals must be processed in parallel, then even the cost of cable becomes an important consideration.

Early efforts at instrumentation involved the use of carrier amplifiers. The bridge structures just discussed were excited at ac. The excitation carrier frequency was about 3 kHz, which of course limited the maximum signal bandwidth. The benefit to this approach was that the signal of interest was in the form of carrier-suppressed modulation. At null there was no carrier. This signal could be transformer-coupled to a single-ended amplifier. The transformer provided ground isolation and resolved many of the common-mode problems. The transformer also matched input impedances and provided low noise levels referred to the input. With a carrier system the noise of interest centers about the carrier frequency, not about dc. The problems were the limited bandwidth and the difficulty of keeping the carrier signals from cross-coupling in cables and causing errors. The error signals were expressed in microvolts, while the excitation level was expressed in volts. Bridge balancing had to be both resistive and reactive, which caused many setup difficulties. This approach was used in a period when there were no good dc amplifiers let alone differential amplifiers.

The carrier system used transformers to couple signals between electrostatic regions (grounds). There are many other techniques available to perform this task. If the signal is digitized, it can be transported using fiber optics. The signals can also be transported over a radio link. In both of these examples the signal is first amplified and conditioned before it is transported. When the signal is amplified first and then transported, the problems are obviously less complicated. Unfortunately this is not always practical. Electronics placed on a structure under test may actually modify the structure mechanically. Often the instrumentation cannot survive the heat or vibration on the structure and must be placed in a blockhouse.

5.13 THE DIFFERENTIAL AMPLIFIER

When a difference signal (the signal of interest) is amplified, conditioned, and delivered to a second ground the instrument is called a *differential amplifier*. An oscilloscope with balanced inputs probably does not qualify as a differential instrument amplifier. The observed signal is the difference between two inputs, but the reference conductor is the framework of the

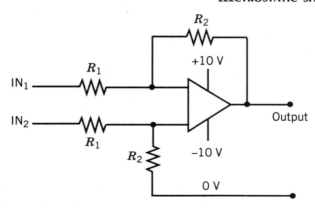

Figure 5.13 A differential amplifier.

oscilloscope which must be connected to the common of the two signals being observed.

An integrated circuit (IC) amplifier has two balanced inputs. Figure 5.13 shows a feedback structure around an IC amplifier that forms a differential amplifier. The differential input impedance is defined by the feedback resistors. This input impedance level might be as high as 100,000 Ω but nowhere near the 1000 MΩ requirement determined earlier. At higher input impedances this circuit has too much noise and drift for it to be practical. See Figure 6.3 for a high-impedance differential input circuit.

IC differential amplifiers are available on the market. The user must provide power, external gain resistors, gain switching, zeroing and any input signal conditioning. The input leads and circuitry must be guarded as described earlier. If the gain range, common-mode level, and bandwidth are acceptable, these devices can serve as instrument differential amplifiers.

This type of differential amplifier can also be used at the source to raise the signal level and provide a low-impedance signal source. This signal can then be carried a long distance with little risk of interference. If the output signal is held below 1-V peak, then post electronics can provide required gain, offset, filtering, and output current. It is practical to provide most of this post-signal conditioning digitally The shielding of the intermediate signal is not critical because the signal output impedance is low and there is no unbalance impedance. The post amplifier should have a differential input stage because of the distances involved and the common-mode potentials that can be encountered.

6

ELECTROSTATIC SHIELDING—II

6.1 SHIELDING LOW-FREQUENCY SIGNALS

Most sensitive analog signals need to be carried in a shielded cable. The shield is tightly coupled to the inner conductors by capacitance. The shield must follow the signal or it will contaminate the signal. Another way to state the problem is to say that the signal must drive the shield. Grounding the shield to a so-called "best ground" or "clean ground" is not correct. If the shield ground differs in potential from the signal ground, then the capacitances in the cable will cause current flow in the unbalance impedance, and this is a contaminating signal. The rule in Section 5.4 must be followed. The shield must connect to the signal where the signal grounds. In this case it would be proper to say that the ground drives both the signal and the shield together. This makes sense because it is impossible to drive or control the ground potential. It might be more correct to say that it is difficult to limit or control the fields that are present in the space between grounds.

It may seem that shielding a low-impedance signal pair is not critical because there is no line unbalance to couple to the interference. Also it may seem that shielding a balanced pair may not seem critical. Theoretically, any coupling that occurs should affect both input conductors equally, and the common-mode signal that results is hopefully rejected. Common-mode rejection usually falls off linearly with frequency. The coupling mechanisms are usually proportional to frequency.

If bandwidth is required, then signal lines may no longer be balanced at high frequencies. This allows an incorrect shield potential to couple interference. The capacitances from the shield to the inner conductors may differ by as much as 50%. This occurs in multiconductor cable when some of the conductors are inside the wire bundle and others are on the outside. This unbalance in capacitance couples more current to one lead than to the other. The high-frequency interference that results is part differential, and this component is amplified as normal signal.

Another problem occurs for active devices. The output impedance rises with frequency as feedback falls off. The signal conductors are thus not balanced, and the shield can couple high-frequency interference into this imbalance. If there is post-signal filtering, this interference may not be observed. A rising output impedance is an inductance, and this inductance is in series with the cable capacitances. If this resonant circuit is not damped correctly, the result can be shield-induced interference of a ringing nature. It is sometimes better to have no shield rather than have one that is improperly connected.

All electronic circuits reach a point where they cannot handle high-frequency interference. Limits occur when currents cannot go negative in a transistor or the required current cannot be supplied to some parasitic capacitance. These limitations cause signal rectification. This rectification results in low-frequency errors that add to the signal of interest. Unfortunately these errors are in-band. This phenomena is known broadly as slew-rate limiting. High-frequency interference most often enters a circuit at its input. Input filtering can remove some of this interference, but there is no guarantee that the filtering is adequate. Input stage filtering is minimal if it is provided at all. It is rarely a part of any performance specification.

In many low frequency instrument designs the shield is not used as one of the signal conductors. In high-frequency transmission the shield is used as a return path (coax). This use of the shield causes trouble at low frequencies for low-level signals. An exception might be a microphone cable. When a quality shield is used and the run is not too long, the interference is minimum. Also note that input microphone cable is not grounded at the source. The hand grounds the shield through a high impedance. If there is an occasional spike of noise, it is not objectionable.

6.2 SHIELDING SINGLE-ENDED SIGNALS

A single-ended signal that is not grounded at the source is best carried on a two-conductor shielded cable. The shield should connect to the signal in

SHIELDING SINGLE-ENDED SIGNALS

the instrument. This connection should keep shield current from flowing in the circuit common. If this shield current causes a voltage drop in the common conductor and this is sensed as a signal, then interference is introduced and perhaps amplified. It is acceptable to terminate the input shield on the conductive enclosure. This will usually avoid any common-impedance coupling. This grounding arrangement is shown in Figure 6.1. The shield should be grounded once. Multiple grounding of the shield is not recommended. Segmenting the shield and connecting these segments to different grounds is also incorrect.

The shield of a grounded single-ended signal must be connected to the signal common where the signal grounds. No other connections are permitted. If the shield is grounded twice, a shield gradient will result and the shield can couple interference into the signal leads. The shield cannot be segmented. If the cable terminates on an intermediate distribution panel, the shield must be carried through with the signal. The distribution panel must not reground the shield. The output signal from the device cannot be grounded because this forms a signal ground loop. The current would flow between grounds in the signal conductor and introduce interference.

This discussion involves best practice. If the signals are high level and the source impedances are low, then improper shielding practices can be incorporated and trouble may not occur. When there is a choice, the best practice should be the rule. The trouble with poor practice is that it can be misleading. If it works, then the practice is repeated. When this bad practice fails, there is no right practice to turn to. If bad practice is followed for economic reasons and the choice is fully understood, then this is good engineering.

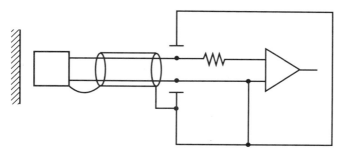

Figure 6.1 The shield connection for a single-ended signal.

6.3 SHIELDING BALANCED SOURCES

A balanced signal source implies equal and opposite signals with respect to a reference conductor. This reference conductor can be grounded or not. A floating bridge can supply a balanced ungrounded signal. One side of the bridge excitation is the reference conductor for the balanced signal. A centertapped transformer might provide a balanced power source where the centertap is the reference conductor. There are logic signals that are balanced in the sense that the logic is always of opposite level on the two signal conductors. The common conductor is the reference conductor, and it is usually grounded at the logic source.

Balanced signals are best handled by differential amplifiers. The difference in potential is amplified, and any common-mode content is rejected. If the signal source is grounded, then the shield should connect to this ground. This shield then becomes the guard shield. The guard shield is not regrounded in the instrument, and it should not be regrounded by the user. Sometimes it may be used in the input circuit as a source of common-mode signal.

If the balanced signal is not source-grounded, then the signal can be amplified using a balanced input stage. If a single-ended input stage is used, then one of the signal leads can be ignored. In effect the second lead provides redundant information. In low-level applications this redundancy improves the signal-to-noise ratio and a balanced input stage is recommended. The input reference conductor should be connected to circuit common. The shield must be connected to the signal preferably to the reference conductor at the signal source. If this connection is not possible, then the shield can be connected to the reference conductor at the amplifier.

Bridge-type input circuits often require one active arm. The strain or temperature or pressure changes the resistance of one element. The remaining bridge completion elements are supplied inside the instrumentation. The active gage element is mounted to the structure under test. This structure should be the grounding point for the signal and the input guard shield. If it is not used as the reference conductor, then capacitances to the structure can add interference to the signal.

When two active arms are used, there are two possible configurations. The active arms can be adjacent or opposite arms of the bridge. The active arms can always be positioned so that any reactive interference is balanced with respect to the input leads. The preferred grounding scheme is to ground one side of the excitation at the structure. This removes the capacitive coupling issue. If this cannot be done, the user relies on the balanced nature of the source and on common-mode rejection for performance. Mounting transducers to balance reactive coupling is chancy at best.

6.4 FLOATING CIRCUITS

Input circuits should not be left floating. Floating means no ohmic connection back to the output common. If the signal source must float, then the amplifier should provide this grounding. A quality differential instrument has high-impedance inputs, and the guard shield is not terminated. This means that this class of instrument normally does not provide for input grounding. In this case a grounding connection should be made by the user.

A floating input circuit provides no return path for small currents that can flow out of an input stage (base current). Many instrument designers will provide a high-impedance path for this current. A typical path might be 100 MΩ. One nanoampere and 100 MΩ causes a signal of 0.1 mV. If this voltage is balanced by an equal voltage on the second input conductor, the resulting offset will be rejected as common-mode signal. If this signal is unbalanced and it is amplified directly, it represents a large signal error.

If the floating input cable develops a leakage path to a common-mode signal such as a dc power supply or a power voltage, the problem of rejecting this signal is left to the instrument. If this signal overloads the amplifier, the performance is obviously compromised. The user may not recognize the problem by observing the output. Basic to all instrumentation is the fact that the answer is an unknown. This is the reason for the measurement. This means that all answers are accepted because there are few alternatives. It behooves the instrumentation engineer to take all of the precautions possible to guarantee the integrity of the results. A slight leakage path to a common-mode source causes no problem if the source is solidly grounded.

When input bias current has no return path to the circuit, the input stage must overload. The overload forces the input stage to operate at some limit point so that this current has a return path. The amplifier appears to function normally (feedback at work), but signal errors are obviously introduced. As stated before, a quality instrument provides leakage paths for current to flow between all input conductors, including the guard shield and the input circuit reference conductor. These paths are usually passive and are about 100 MΩ. This requirement is seldom mentioned in a specification. These leakage paths allow the instruments to perform even when the application violates the grounding rules. In some designs an input-to-output path is provided by a precision balanced attenuator located between input and output circuits. In this type of design there should still be leakage paths between input leads and the input guard. In some designs a leakage path is provided between input guard and output common. This path might be 1 MΩ. Allowing a few microamperes to flow between grounds poses no threat.

6.5 SHIELDING GROUNDED UNBALANCED CIRCUITS

Examples of unbalanced grounded sources include thermocouples, signals from generators, output signals from amplifiers, and some crystal transducers. When the signal levels are high and the source impedance is low, shielding may not be critical. Output conductors can be coax, shielded twisted pairs, or even ribbon cable. Any shield should connect to the source ground if practical.

Thermocouples deserve special attention. If the thermocouple floats as in the case of a fluid temperature measurement, there is no source ground. The only ground is through a capacitance to any nearby structure. Thermocouples must be associated with a thermal reference junction that acts as a heat sink. This reference junction is held to a fixed temperature. The conductors after the reference junction are simply copper. The reference junction is associated with another ground that must be shared with all signals using the reference. Some reference junctions have provisions for individual guard shields. This reference junction ground is not connected to the signals, but it is a source of interference from capacitive coupling.

In general, thermocouples have a few-hertz bandwidth. This means that a filter can be used to eliminate most coupled interference. In most instruments this filtering occurs at the output of the instrument. If the unfiltered signal is noisy, then there is a chance of signal rectification. In this circumstance the signal is compromised. One way to resolve this problem is to allow plenty of signal "headroom." That is, operate at 1 V peak full scale instead of 10 V peak. Another way to stop this overload is to prefilter the signal. This can be done in a signal conditioning unit at the input to the amplifier. If post filtering is used, it is wise to observe the unfiltered signal to make sure the noise and interference are within the limits of the amplifier and of the filter during the test.

Floating thermocouple signal leads should be grounded at the reference junction or at the instrument. The thermocouple signal can be balanced and filtered by introducing a slight attenuation. Figure 6.2 shows this circuitry added into the input leads at the input to the instrument.

SHIELDING ELECTRONIC EQUIPMENT

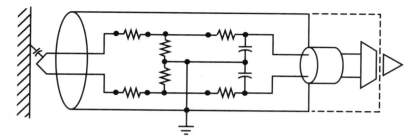

Figure 6.2 Thermocouple balancing and filtering.

Bonded thermocouples are necessary to obtain reliable temperature information. Any insulation can form a temperature divider, and this results in an error. When fast temperature changes are expected, then the thermocouple must be welded to the point of measurement. In blast studies or in impact studies, very rapid temperature changes are expected.

A grounded unbalanced source should be conditioned by a differential instrument amplifier. The input signal should be guarded to limit interference. The guard shield can connect to the structure or, if necessary, to the signal common at the reference junction.

Grounding crystal transducers is a separate topic covered in Section 8.6.

Grounded unbalanced signal sources are usually high level. An example might be a signal generator with a 50-Ω output impedance. For most bench work it is acceptable to use a simple coaxial cable to connect generator signals. For long runs it may be preferable to use a two-conductor shielded cable with the shield source grounded.

6.6 SHIELDING ELECTRONIC EQUIPMENT

The conductive enclosure for a bench-operated instrument should be connected to equipment or safety ground. This is for the safety of the user. The power cord has a third pin that provides this connection. Equipment is designed this way to meet various code requirements. If the external case is plastic and the controls are insulated, then this safety issue is removed.

The earth ground provided by the equipment ground conductor (green wire) is not the correct shield potential for signals brought to the equipment. This means that the designer may have to insulate input connectors and provide separate internal signal shields. If the equipment generates a signal and the case is connected to one of the signal conductors, then the signal reference conductor is tied to the equipment ground. This too is not usually an acceptable approach because it forces a specific grounding when the signal is connected to a circuit. The solution provided on many pieces of equipment is a grounding jumper. This jumper can be easily removed and allows the circuit to float from the enclosure. The circuit is grounded by the user when the signal is externally connected.

It is easy to see that there is a problem when several pieces of equipment are interconnected on a test bench. Each piece of equipment is grounded separately. Even if the internal circuitry floats from the conductive housing, the transformers are a source of interference. In every device the primary voltage is in series with transformer capacitances, and this introduces current flow in signal conductors. Fortunately, on most test benches the lead lengths are short and the signal levels are high. This current is interference and it has high frequency content. If the same setup is used where cable runs are long, the resulting interference may be objectionable.

The grounded oscilloscope causes its share of problems. Most oscilloscope users break the safety connection by using a "cheater plug." This way the oscilloscope does not connect the circuit being monitored to the safety ground. Of course the power transformer still provides a reactive path to this ground. The only way to avoid this path is with a battery-operated oscilloscope. With the "cheater plug" installed the oscilloscope could be connected to the hot side of the power line, making the housing a shock hazard. There is a temptation to do this when off-line switchers are being tested or designed.

When the oscilloscope must be referenced to either side of the power line, it is preferable to use a power isolation transformer to operate the circuit. The oscilloscope then grounds that side of the power line it is connected to. In this manner the test equipment is always safe. A 120-V/ 120-V transformer rated at 1 kW will usually suffice. No shields are required. This transformer cannot be a part of the facility wiring because it does not meet code requirements. The code requires that one side of the secondary be grounded to the nearest point on the grounding electrode system for the facility. At the end of the test the transformer should be disconnected.

6.7 SHIELDING INSTRUMENTATION AMPLIFIERS

The conductive housing for an instrument amplifier is usually connected to the guard shield. When the unit is installed in a cabinet, this housing is insulated from the cabinet. This treatment makes it practical to provide

SHIELDING INSTRUMENTATION AMPLIFIERS

extensive input signal conditioning without adding to the leakage capacitance out of the input guard shield. The total leakage capacitance must be held to under 2 pF.

The potential between the guard shield and the output common is usually only a few volts. In applications involving long input leads, ground potential differences can momentarily be hundreds of volts. The shock hazard is minimum. In a good design the front panel metal is connected to output ground to avoid any chance of shock hazard. In instruments capable of handling 300 V of common-mode signal, an inadvertent connection to a high voltage might go unnoticed. A user coming in contact with this voltage would be shocked.

The guard shield forms a transmission line with other external conductors. Currents from these external signals must flow in the shield. This current flow changes the guard potential as measured at the instrument. It is preferred not to use the guard shield to drive internal instrument circuitry. Any loading would further modify the guard potential and affect the performance of the instrument as a system. For short input leads, this problem would not be apparent. For long input lead lengths (greater than 300 feet), this loading could start to be an issue. If the guard shield were used to drive internal circuitry, any interference on the transmission line would enter the circuitry and degrade the performance. It is not practical to filter the guard shield potential because there is no reference conductor available.

An accepted practice is to regenerate the common-mode potential from the signal leads. One solution is to sum the two signal leads through 100k resistors and buffer this sum with a FET follower. In an ideal situation this voltage should equal the voltage on the guard shield. The resulting signal is the common-mode voltage free from the interference that uses the signal cable shield as one side of a transmission line. The power for the FET follower is referenced to the output common. There is no filter in this circuit, and there is very little phase shift in the FET follower. The output impedance of the FET follower is low compared with the undefined source impedance of an uncontrolled transmission line which describes the guard shield. The FET follower can be used to drive internal power supplies that help raise the input impedance. This circuitry is shown in Figure 6.3. The derived common-mode voltage is offset by the gate-to-emitter voltage drop. This in no way reduces the performance of this circuit. A high input impedance requires that the internal power supplies follow the common-mode level dynamically. These changes are accurately provided by this circuit.

Local shielding of sensitive circuits is a common practice. A small plate of metal or a small metal box can be used to cover or surround a circuit.

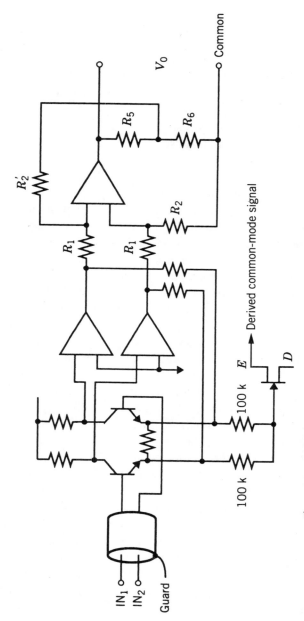

Figure 6.3 A high-impedance differential amplifier showing a derived common-mode signal.

THE DRIVEN SHIELD

In charge amplifier designs, some of the leakage capacitances must be held to below 0.001 pF and a metal box is the only practical solution. The box or plate is connected to the local common, not to the input shield. In a sense the shield is segmented. This segmenting idea is not allowed on external signal cables.

Ground planes on printed circuit boards can be used as limited shields. In most analog designs a two-layered board is used and a ground plane is not available. In digital designs that mix analog and digital signals the story is different. The ground plane functions as a capacitive divider and reduces coupling. The components mounted to the board are not enclosed by the ground plane, and thus there is no protection from nearby cables or nearby components. Circuitry mounted on a neighboring circuit board can also contribute interference. A plate mounted above the components can serve to eliminate this type of coupling. In all circuits, ventilation is important and shields cannot impede the flow of air. The issue of separate ground planes for analog and digital circuits is discussed in Section 10.16.

6.8 THE DRIVEN SHIELD

A good example of a driven shield was given in the previous section. The derived common-mode signal could be placed on a metal plate and used as a driven guard shield. Usually the problem is one of eliminating capacitances rather than controlling them. For some high-impedance sources a shielded cable attenuates the signal of interest. The cable capacitance forms a simple RC filter and the frequency response is limited. The idea is to drive the shield with the signal to reduce the shield capacitance. Up to a point this can be done. This driven shield is a complex feedback circuit with transmission line delays. This kind of circuit can easily become unstable. Each application needs to be adjusted individually to make sure that there is no chance of instability. This may be impractical when the cable lengths are undefined. A typical driven shield circuit is shown in Figure 6.4. The difficulty with this circuit can be appreciated if a step input signal is applied. The voltage that arrives at the amplifier is small. The correcting signal that returns on the inner shield is also small. A new, higher voltage is now sent forward to the amplifier, and again the voltage on the inner shield is increased. This incremental process continues until the output voltage equals the input voltage. The charge in the input cable is eventually supplied by the amplifier rather than by the source. If the amplifier has some gain, then the charging process can be accelerated. If the gain is too high, then

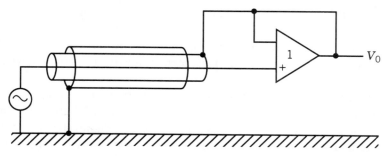

Figure 6.4 A driven shield circuit.

the system is an oscillator. The maximum acceptable gain is a function of cable capacitance and cable length. Tests should always be made to determine that the damping factor is within reasonable limits.

6.9 THE ANALOG INPUT CABLE

Many different shielded signal cables are available on the market. Many of these cables are designed to handle high-frequency signals. The manufacturer provides the characteristic impedance, and in some cases it provides transfer impedance data (see Section 10.11). For analog applications these data are usually unimportant. The unterminated frequency response as a function of cable length would be nice information to have, but it is rarely supplied.

The shields on many cable types are braid (woven copper strands). This braid provides good cable flexibility. The braid can be tightly or loosely woven, and this affects price as well as performance. In some cables a double braid is provided. There is an improvement in high-frequency performance (above 1 MHz) if the two braids are insulated from each other. At low frequencies two braids improve the optical coverage of the inner conductors, and this results in a lower leakage capacitance (lower mutual capacitance) to nearby conductors.

Twisting of signal conductors is standard practice in many analog signal cables. This greatly reduces differential coupling to fields that can enter the cable. At each half-twist the induced voltage cancels. The fields that cause the most trouble for low-level instrumentation signals are near induction fields. These are usually magnetic fields from power-related circuits.

THE DRAIN WIRE

Copper braid or aluminum shielding is ineffective against this type of field. Two-conductor twisted shielded cable is very commonly used to handle low-level analog-type signals. Multiconductor (up to 10 leads) shielded cables are used for strain-gage work or general instrumentation. Groups of multiconductor shielded cables can be supplied in one larger cable. All of the shields are insulated from each other for the run of the cable. This type of cable is available on special order.

6.10 ALUMINUM FOIL SHIELDING

A question often asked is whether aluminum makes a good shield. The answer is yes. There are a few problems associated with aluminum foil shields when used in a cable. It is difficult to make a soldered connection to aluminum. For this reason, cable manufacturers provide a drain wire that makes contact with the aluminum foil along the cable run. The drain wire is used for a shield connections at cable ends. The foil is usually one or two mils thick to provide flexibility, and it is very fragile. An outer insulating jacket protects the foil. At connectors or at points of flexure the shield is still apt to split. The outer surface of the aluminum is usually anodized to limit corrosion and extend the life of the cable. The foil is wrapped around the cable so that a seam follows the length of the cable. Along the cable run the leakage capacitance is extremely small.

Aluminum foil shielding does not satisfy the requirements for a good transmission line cable. The seam, for example, does not allow the free flow of current on the inside surface. The lack of conductor symmetry, the unsymmetrical drain wire, and the twisting of the conductors makes this a poor high-frequency cable. However, for a run of a few feet the cable may be entirely satisfactory even at 100 MHz.

6.11 THE DRAIN WIRE

For analog applications, aluminum foil shielding can be quite adequate. There are problems when the cable is used in regions of large radio-frequency (rf) fields. In aircraft production, large rf welders are often in use. Testing in this area may be impacted by these fields. Aluminum foil shields are not advised in this application. The problem relates to the drain wire. The drain wire carries some of the coupled rf current to the inside of the cable. This, in turn, couples to the signal conductors and adds to the common-mode signals at the input to the instrumentation. This, in turn, adds to the chances of signal rectification and signal error. Note again that filters at the input to most instruments are minimal, if present at all.

In applications involving analog and digital signals in one shielded cable, an internal drain wire in a cable can pose problems. This drain wire might be used to connect cabinets or enclosures together. The cable shield should be used to carry interference signals on its outside surface. If the shield is not adequately bonded at its ends, then the easiest path for interference currents to follow is the drain wire. This drain wire couples interference into all internal conductors, including logic leads. This interference can cause digital errors or impact the quality of the analog signals. In industrial applications where solenoids and relays abound, the interference levels can be high. A poorly bonded shield will work better than a shielded cable with an added drain wire. The drain wire couples field to the cable along the entire cable length. A poor shield bond allows field energy to couple to the signal conductors at the ends of the cable.

If a drain wire is provided on the outside of the foil shield, this is an improvement. The outside surface of the aluminum must be conductive or the drain wire will be ineffective.

6.12 LOW-NOISE CABLE

In some applications, special low-noise cable is required. Applications include very high impedance signal sources and cabling to piezoelectric transducers. Special coatings are applied to the conductors and insulators to limit charge buildup. The braid is usually one of the signal conductors. Charges are generated on dielectric surfaces when the cable is moved. This is known as a *turboelectric effect*. Low-noise cable also limits pyroelectric effects. This is the generation of charge when the cable undergoes a temperature change. If this charge is generated in the transducer, then the cable cannot reduce the effect.

Long runs of low-noise cable may not be an acceptable choice. The shield is a signal conductor and can couple to interference. This type of conductor is also expensive. Piezoresistive elements can be used in a bridge circuit to measure vibration. These elements are of nominal resistance and are compatible with standard cables. It is difficult to maintain bridge balance, but fortunately this is compensated for by higher signal levels.

6.13 REACTIVE COUPLING IN CIRCUITS

In many circuit layouts, the coupling between circuits or components can be treated as a simple capacitance or inductance. Reducing capacitive coupling involves increasing the spacing, reducing the victim circuit impedance, or interposing a shield. The shield is connected to the victim circuit common and this forms a capacitive attenuator. Traces connected to a circuit common that surround a critical trace can also be effective in providing this shielding.

When an interfering signal capacitively couples to a group of conductors it can introduce a common-mode signal to that group. The reference conductor in this case is the signal common for the culprit. Capacitive coupling can introduce both normal and common-mode interference. If the interfering signal is normal mode then the next issue is whether it is in-band or out of band. In either case if the interference causes slew-rate limiting then an in-band error can result. Obviously this coupling should be eliminated. If the coupling is common-mode in nature then the question of common-mode rejection is important. A common-mode signal that is not rejected results in a normal mode signal.

The degree of capacitive signal coupling depends on victim circuit impedance and circuit balance. The source impedance of the interfering signal can usually be ignored. This is because the coupling impedance in series with the source is high (a small capacitance). Keeping victim circuit impedances low reduces the degree of capacitive interference from a nearby source. High input impedance devices are very vulnerable until they are connected to a reasonable source impedance.

If the interference is inductive in nature then the impedance of the interfering source is important. A lower impedance source will allow more current to flow; thus involving a greater magnetic field.

Inductive coupling can also introduce both normal and common-mode interference. If the field couples to an entire signal group then the effect is common-mode in nature. If the field couples to the space between signal leads then the interference is normal mode in nature. Again the issues of in-band, out of band, normal mode, common-mode, slew-rate limiting and common-mode rejection must be considered. It is important to realize that a common-mode signal is an average signal and that a part of this signal may be normal mode. Thus one signal may cause two modes of interference.

The capacitances that result from a given circuit geometry can be difficult to calculate. The best way to treat this problem is to consider the simplest possible geometry. The capacitances between parallel conductors is given in many textbooks. Similarly the fields that couple between circuit loops is a problem in mutual inductance. Equations for mutual inductance are also in the textbooks. For inductive coupling it may be easier to calculate the B field and solve for induced voltage. The resulting current or voltage in the system of interest can be used to calculate the interference.

Inductive coupling that causes differential interference can be reduced by reducing loop area. For example, signal paths in low level circuits should have very small circuit loop areas. The simple twisting of conductors in a cable can reduce differential coupling. This twisting is ineffective in reducing common-mode coupling, which can be reduced by limiting the loop area that allows this coupling to occur. For example, cables should be routed on ground planes. The other methods available are to increase the culprit-victim spacing or raise the source impedance to limit the interfering current.

There are several ways to reduce the magnetic field that radiates from an interfering (culprit) conductor pair. The best way is to run the pair as coax. The next best solution is to twist the two conductors. Twisting of traces on a pc board is not practical. Circuit geometries that can be quite effective in reducing the external magnetic field are available. A quad of conductors with the diagonally opposite conductors used as one conductor has a very low external magnetic field. Two broad traces very closely spaced will also have a small external field.

Before making changes, the interference must first be characterized. If an added shield reduces the error then the coupling is probably capacitive. If the orientation of leads changes the coupling then the coupling may be inductive. In some cases the experiment may have to be more sophisticated. Raising or lowering the culprit current level leaving the voltages alone can be used as a good indicator. If the interference is proportional to current, it is inductive.

In a few cases the coupling may be both capacitive and inductive. The suggestions made above still apply but any reduction in interference must involve both modes of coupling.

6.14 GUARD RINGS

In some high impedance circuits, the resistance of the pc board itself limits circuit performance. For example, currents flowing in this resistance can enter the circuit at a summing point. One solution involves a ground ring placed around the summing point. Currents now flow to this ring rather than enter the inside of the ring and flow to the circuit. This approach can sometimes be far less costly than mounting the circuit on a ceramic board. Of course, the guard ring may be required on both sides of the board.

7

COMMON-IMPEDANCE COUPLING

7.1 INTRODUCTION

There are many circuit arrangements that allow signals to cross-couple. The coupling may involve signals of interest or interference. The coupling is often interrelated with signal conductor grounding or signal isolation. Many of these problems can be described by using circuit theory rather than field theory.

It is easy to see why semantics enters into the problem. An isolated output signal is a good sounding phrase, but the question arises as to what the output is isolated from. In most applications the desire is to connect a single signal to two different devices. The two output signals might share the same common but arise from separate buffers. This is one form of isolation. If one output shorts out, the other will still function. When the outputs are connected to two different grounds a ground loop is formed. Another form of isolation might imply separate signal commons, but this is rarely supplied.

Every conductor is an impedance. It has inductance and resistance per unit length. When unwanted current flows in a signal conductor, the impedance is shared with the signal and the result is interference. This is the common impedance coupling that is referred to in this chapter. If the common impedance is a part of a transducer, then the problem is obviously more severe. In Chapter 5 the discussion centered around keeping power currents flowing in shields away from signal conductors. The flow of any unwanted current in a signal conductor results in common-impedance coupling.

7.2 AMPLIFIER COMMON IMPEDANCE

The power to an electronic circuit should enter at the output stage of the device. This forces output signal current to flow directly to the power supply. If output signal current flows in an impedance associated with the input, then current feedback results. In high-gain circuits this can lead to instability and gain error.

In single-ended circuits the input and output commons should not both be grounded. Aside from the ground loop that is formed, the grounding provides a second path for output current to return to the power supply. This path involves the common lead that enters the circuit at the input. This is a form of current feedback. The effect is more pronounced when input leads are long. The lead impedance provides significant common-impedance coupling. High-gain amplifiers can be unstable as a result of this feedback. This feedback affects the gain, which changes as a function of load. It is possible to build negative-impedance amplifiers this way. As the load resistance decreases, the voltage gain rises. The two current paths for output current flow are shown in Figure 7.1.

Several electronic devices can share the same power supply. If output signals from these circuits are multiply grounded, then signal currents can take several paths to return to the power supply. This may not cause gain errors, but there can be crosstalk at high frequencies. In audio work where all of the signals result from the same audio source, some crosstalk can be tolerated. When the signals must remain separated at all frequencies, this crosstalk is not acceptable.

7.3 STRAIN-GAGE EXCITATION SUPPLIES

In strain-gage instrumentation the cost of individual excitation supplies is high. There is often pressure to avoid this approach. Here are the arguments for individual supplies. A common supply can limit both reliability and performance. In a common supply, individually grounded gages form ground loops with the excitation common. Current flowing in these ground

STRAIN-GAGE EXCITATION SUPPLIES

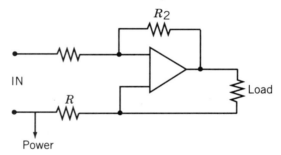

Load current returns through R, and this is feedback.

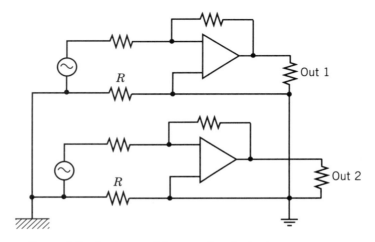

Some of the load current from each output returns through the grounds and uses resistor R, this is feedback.

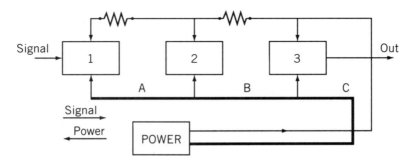

Signal currents in cicuit 3 do not flow in impedance A

Figure 7.1 Common-impedance feedback paths.

loops cause input common-mode signals. The balanced nature of the inputs will reject this common-mode voltage. However, when the bridge is unbalanced by a normal signal the interference will enter the signal path as signal-modulated noise.

If the excitation supply is grounded once at one of the gages, then this ground is not correct for the other gages. The parasitic problems that result are shown in Figure 7.2. This problem is most pronounced when the bridges have only one active gage element. The source is unbalanced with respect to the interference. Ground potential differences cause interference currents to flow in the transducer impedance, and this generates a normal signal. A transducer may have a resistance of 500 Ω, not the few ohms

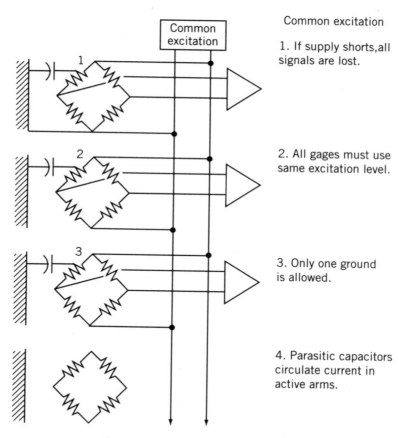

Figure 7.2 The single excitation problem.

STRAIN-GAGE EXCITATION SUPPLIES

found in an input conductor. With two active arms the coupling in the second arm can cancel or be additive depending on the bridge configuration used. In any case it is unwise to expect that these capacitances will be equal or that a parasitic capacitance will couple noise current to the gage element in a predictable manner.

With separate supplies, excitation voltages can be set to accommodate a variety of gage types. Individual excitation sources allow for remote sensing at each gage. Individual excitation sources allow one input guard shield to function for each signal conditioner. Perhaps the most important reason for individual excitation supplies involves reliability. Often a test costs more than the instrumentation. If one gage shorts out and the excitation source fails, all of the data are lost.

If the gage excitation is supplied constant current, then there can only be one excitation source per gage. This mode of excitation provides an improvement in signal linearity for many bridge configurations. A word of caution. Bridge balancing using a shunting potentiometer and current injection violates the constant current source impedance. Series voltage insertion can be used for bridge balance without affecting the source impedance. Typical output impedances for constant current supplies are 10 MΩ. A typical source impedance for voltage supplies is 0.01 Ω. A measure of output impedance must be made at or near the output connector.

Problems exist when maintaining the expected excitation impedance at the gage when long cables are involved. The 10-MΩ output impedance of a constant current source is developed by using feedback. The feedback factor must fall off as a function of frequency or the system cannot be stable. This reduction in feedback at high frequencies makes the output impedance appear as a shunt resistance and capacitance. If the impedance is 100 kΩ at 10 kHz, the effective shunt capacitance is 159 pF. It is easy to see that any input cable adds to this shunt capacitance. A typical cable adds 30 pF/ft. Obviously the linearity expected out of the bridge cannot extend to high frequencies because the excitation supply can no longer be considered constant current. This may not be a problem if the expected signal levels are small at high frequencies.

The low output impedance of an excitation supply cannot be maintained at the gage. This is because remote sensing has limited bandwidth. The excitation source impedance above a few hundred hertz will be that of the transmission line. A transmission line connected to a zero-impedance source looks like a dc resistance in series with an inductance. Fortunately, for short cable runs this does not introduce a significant error. If remote sensing is not used, the excitation impedance at the bridge is the lead resistance

116 COMMON-IMPEDANCE COUPLING

plus the inductance in the cable run. If the temperature varies 40°C, the cable resistance can change from, say, 5 to 5.8 Ω. For a 100-Ω bridge this is about a 1% error. This is the reason why remote sensing is needed on long cable runs. The sensing current can be in microamperes. The remote sense voltage must be buffered by a differential amplifier. If this is not done, the remote voltage will not be correct.

7.4 TRANSFORMER SHIELDING FOR EXCITATION SUPPLIES

The excitation supply used for strain-gage instrumentation must be inside the input guard shield environment. This region normally has no other active circuits (see Section 5.12). The power transformer ideally should have three shields. Two shields are acceptable with careful coil construction. This compromise assumes that one side of the excitation is grounded at the transducer and that the signal leads are balanced. The primary shield is connected to safety ground and the second shield is connected to the signal guard. This shielding configuration is shown in Figure 7.3.

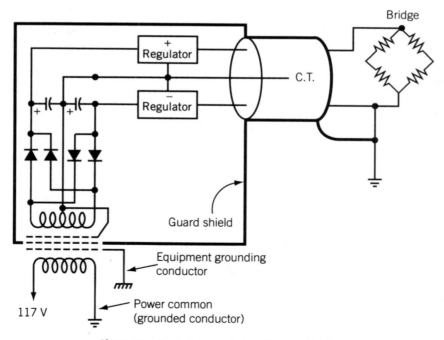

Figure 7.3 Excitation supply transformer shielding.

The transformer secondary voltage can circulate current in the excitation common. To limit this current and avoid a third shield the secondary coils can be configured so that the outer layers of the coil form the centertap. The voltages in series with the capacitance to the guard shield from the secondary coil can be reduced to about 1 V instead of 30 V. If all the coil leads are brought to the outside of the transformer, then the user can form the centertap. If the outer layers of the coil form the centertap, then the voltage in series with the secondary capacitance is reduced. In effect, the higher voltages are shielded by the turns of the coil.

If it were necessary to ground the signal and float the excitation, then a third shield in the power transformer would be required. The unbalance impedance now becomes 1000 Ω. To keep the interference coupling below 1 μV, the current in the unbalanced impedance must be less than 1 nA. Assume the driving voltage is 30 V at 60 Hz. The leakage impedance must be greater than 3×10^{-10} Ω. This means that the leakage capacitance out of the secondary shield would be limited to 0.08 pF, a very small number. If the centertap is carefully constructed, the series voltage is reduced and the leakage capacitance can be as high as 2.4 pF. This is still an expensive transformer.

7.5 STAR CONNECTIONS

In the period around 1950, it was standard practice to use star ground connections in instruments. A central point was selected for the ground of the instrument. A long stud at this point connected all of the grounds of the instrument together. These conductors included the chassis, the input shield, the output shield, the transformer shield, the power common, and the current return conductors from each stage of gain. The idea was simple enough. Current in any one stage returned to the power supply and did not share a common impedance with any other signal ground. Within the dimensions of one instrument and for bandwidths that did not exceed 20 kHz, this concept worked well. Designers even specified the order in which connections were made to the stud. This arrangement of common conductors is also known as *single-point grounding*.

Star connections are used in power distribution systems. Facility power is brought in to a service entrance. Feeder lines leaving the service entrance form a star. The neutral or grounded side of the power is earthed at this point. The neutral or grounded conductor may not be connected to earth at any other point in the facility. All equipment or safety grounds are also brought back to this same point. This arrangement of conductors is fundamental to fault protection. If power current can return over a second path, the fault protection system may not function.

Power can be supplied from a separately derived system. This might be a distribution transformer or an auxiliary generator. These power sources are treated just like a new service entrance, and the new neutral or grounded conductor can be earthed. Power supplied from this source forms a new star connection. The earthing of the separately derived system must be to the nearest point on the grounding electrode system for the facility. A separate isolated earth conductor would be illegal and very unsafe.

The neutral that is formed in the new star connection supplies current to the new load. Neutral voltage drops in other parts of the facility do not affect this neutral. This is the technique used to supply power to a computer facility where the quality of power is critical.

The star connection fails when it is extended to digital circuits, multiple instruments, or several buildings. The loop areas that are formed are large, and many new problems arise. As an example the star grounding of signals arising from different buildings means long grounding conductors. The potential differences in the earth when a lightning strike occurs can exceed 10,000 V/m. This voltage, if impressed across the coils of a transformer, can destroy the transformer. This occurs when signals are grounded at a remote grounding well and the transformer primaries are grounded at the service entrance. Long heavy copper conductors do not limit this potential difference. Just remember that long conductors are inductors, not short circuits. A lightning pulse is characterized at a frequency of 640 kHz.

Rules applied to power grounding have been acquired the hard way, namely through experience. Rules applied to electronics can also work, provided that they are the right rules. Relying on rules when they are not applicable can lead to disaster. It is human nature to look for rules because this is easier than thinking the entire problem through. Rules can be formulated to help a designer in a limited sense. The designer must recognize when the rules are not applicable. Extending the star connection concept to systems of any significant dimension or to any frequency is obviously wrong.

Within a single rack the star concept can be effective. When signal cables crisscross between many devices the resulting loop areas can cause problems. If a central point is selected and all cables are routed past this one point, the loop areas are controlled. This cabling arrangement can sometimes result in a quiet system. A typical application might be a rack of audio equipment.

7.6 NEUTRALS AS COMMON IMPEDANCES

When there is one distribution transformer in a facility the neutral or grounded conductors must provide the current return path for all loads. If two feeders are supplied from the service entrance, then the there are two return paths and they do not share a common impedance. Running from a separate feeder can be an effective way of limiting some power-related interference.

The neutral (grounded conductor) is brought into every power transformer used in a facility. Voltage drops on this conductor are the result of load currents. This voltage drop appears between the neutral (grounded conductor) and the earth at the service entrance. This voltage is a common-mode voltage where the earth is the reference conductor. The equipment grounding conductor does not carry load current, but it does carry current from filters located in most devices. The equipment ground conductor also services every device. This filtered current must return to the service entrance to complete the circuit loop. This means that the potential to earth from the neutral (grounded conductor) and the equipment grounding conductor are not the same. They serve different functions. The voltages depend on the power distribution system, the number of feeders, the types of loads, and the distances involved. In large facilities these potential differences in every piece of equipment can be a serious source of signal contamination.

The neutral conductor and the equipment ground conductors should be routed together, preferably in conduit. The fields associated with these interference signals should be confined to the conduit. If the conduit path is broken or the conductors take separate paths, the results are fields that extend into the facility. The code requires that the conduit be continuous and the equipment grounding conductor be run in the same conduit. Conduit helps to confine the fields, but they cannot be kept out of the transformers. Transformers provide a reactive direct path into all signal conductors and signal shields. Shorting the equipment grounding conductor to the neutral (grounded conductor) is illegal.

Neutral currents in three-phase systems are near zero for balanced linear loads. In electronic circuits where there are rectifiers with capacitance input filters, switching regulators, and SCR switched controllers the load currents are not linear. If these loads dominate, the neutral currents can be higher than the phase currents. In some facilities, excess neutral current have actually burned out the neutral conductor. This is a serious problem because the code requires all phase conductors to be the same size. The code also does not allow trimming oversize conductors for connection to electrical equipment. This means that the heavy neutral load requires both oversized conductors and larger breakers for all conductors.

When three-phase power is transported using a delta/wye configuration, there is no neutral conductor on the delta side. On the wye side the neutral is regrounded per code and the neutral current is under control. This is the principle used to limit the neutral common-mode voltage that affects every piece of equipment. This power configuration is used in so-called clean computer power systems.

7.7 THE EARTH AS A COMMON IMPEDANCE

The key issue in power distribution is lightning protection. Transmission lines that are above ground are protected by overhead ground wires, and many earth connections to the neutral are provided. The intent is to keep lightning current from entering equipment and burning out transformers. Multiple grounding of the neutral implies that some neutral current will use the earth path. If there is metal conduit in the earth, this becomes the preferred path. In an industrial area the earth is teaming with metal conduit. These conductors bring ground currents into each facility even if the power for that facility is disconnected.

Soil impedance varies significantly. In areas that receive rain, the resistivity of the soil can be 1000 Ω-cm. In the desert the sand may be close to an ideal insulator. Facilities located on a granite slab or on a lava field may be totally insulated from earth. The tundra varies depending on the depth of the freeze. In some areas there are layers of conducting soil between insulators. Obtaining an earth connection is not always guaranteed. In situations like this, all above-ground conductors must be bonded together.

Large cross sections of the earth represent low impedances. The resistance across a cubic meter of earth where the soil resistivity is 1000 Ω-cm is 10 Ω. Across a 10-m cube the resistance is 1 Ω. This assumes that current flows uniformly in the volume of earth. When a connection to earth is made, the current must concentrate around the conductors in contact with the soil. This concentration of current raises the voltage gradient near the conductors. As a result, the resistance to earth can rarely be made less than 1 Ω. This is a low frequency value that does not include the magnetic field energy (inductance) storage associated with this current flow. The NEC requires a 20-Ω connection at each service entrance. If this is not

THE EARTH AS A COMMON IMPEDANCE

achieved, then two grounding connections must be supplied and no more questions are asked.

The earth is a source of infinite charge of either polarity. To say that the earth is a reference conductor of low impedance is true in the broader sense. Locally it is high impedance compared with a metallic ground plane. An earth connection might be 10 Ω and a connection to a 1-mil-thick sheet of copper may be 100 $\mu\Omega$, for a ratio of 10^5.

For commercial radio transmission the ground plane is an adequate conducting surface if a counterpoise is installed. Without the counterpoise the virtual earth can move down depending on how dry the soil becomes. Frequently one skin depth is used as a measure of this depth. The skin depth depends on both frequency and conductivity.

When lightning hits the earth the current spreads out radially. Within a few hundred meters of the strike point the voltage gradient can be lethal. It is not uncommon for cattle to be electrocuted near a strike when they are standing on wet ground. Obviously these high voltages must exist between buildings. If signals must interconnect between buildings, simple earth grounding connections cannot limit the voltage. Because the characteristic frequency of a lightning pulse is 640 kHz the impedance of any conductor connecting the two grounds together will also be inadequate. Protection for electronics must be provided in the form of gaps and surge suppressors. In both buildings the circuits should be placed in conduit, and the conduit should be bonded to the grounding electrode system.

Ground planes could in theory connect two facilities together. The dimensions of the connecting grid would have to be greater than the size of the two facilities. Every earth point in each facility would make contact with the grid. This approach is usually deemed impractical. It would be practical today to communicate between facilities using fiber optics. This solves the lightning control problem very nicely.

Sunspot activity causes large magnetic fields to reach the earth. These signals have frequency content well below 1 Hz. In large conductive loops the induced currents can be significant. For example, if a metal oil pipe line is supported on steel pilings set into the earth, the loop area can be large. In the tundra the soil might be conductive at a depth of 50 ft. Currents as large as 400 A have been observed. Large conductive loops exist when utilities multiply ground neutrals. If this current flow trips a ground fault detector, then power is disrupted. If this occurs in several locales at the same time, the utility may be forced to drop the entire power grid.

A wet soil is an electrolyte. Different metals that penetrate the earth form a battery. If there is current flow, then electrolysis can take place.

This electrolysis can eventually destroy a building. This problem is severe when the earth is used as a dc return path for power supplied from the facility. Two examples are power for electric trains and telephone circuits. If return current is collected on a ground ring(s) around the facility, the building can be protected. Current flow should be monitored to confirm that very little returns in the building steel. If required, a countervoltage can be applied between the building steel and the ground ring to further limit the current flow in the building steel. This voltage might be the output of a feedback system that nulls current flow.

Boats at anchor form a battery between the earth and the hull. This results in hull corrosion unless a sacrificial anode is placed into the water. The standard approach is to use a block of zinc. The zinc is connected to the hull. The zinc is lower in the electromotive series and enters the electrolyte first.

7.8 THE FORWARD REFERENCING AMPLIFIER

When high-level signals (above 1 V) are coupled between equipments there is the standard ground potential difference. This ground potential difference is a common-mode signal. This unwanted signal can usually not be reduced by adding grounding conductors. A simple circuit can serve to couple the signal between the grounds at unity gain. Common-mode signals of a few volts can be accommodated. The common-mode rejection ratio (CMRR) can be in excess of 80 dB.

The basic circuit is shown in Figure 7.4. When the resistor ratios R_1/R_2 equal R_3/R_4 the differential gain is R_2/R_1. If lead ① is grounded, the gain is positive. If lead ② is grounded, the gain is negative. Two adjustment potentiometers are shown. R_5 adjusts gain and R_6 adjusts the common-mode rejection. These adjustments are only needed if there is no other way to set the gain. These two adjustments are not independent, and several adjustment passes may be required. If 0.1% metal film resistors are used, the gain and CMRR will often be quite adequate.

The power for this circuit must be supplied by the output circuitry. A regulated ±15-V supply will suffice. Multiple circuits can be operated from the same supply. The circuit is inexpensive because four amplifiers can be supplied on one IC chip. The resistors are only a few cents. The most expensive parts for this circuit are the connectors.

The author has used this form of isolator for video signals. When video signals are patched, there are often ground potential differences. These potential differences add noise to the video signal and impact frame syn-

DRIVEN POWER STAGES (ISOLATED OUTPUT)

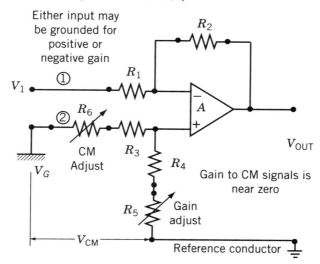

Figure 7.4 A forward referencing amplifier.

chronization. The feedback resistors should be around 3 kΩ. The IC amplifier must be selected to have sufficient bandwidth. Reasonable CMRRs can be obtained out to 10 MHz.

Ground planes used on printed circuit boards have a very low ohms-per-square value (see Section 4.2). The voltage gradients on the board still depend on current density. Near power or signal connections the voltage gradient is apt to be the highest. In some circuit applications a few milivolts of potential difference can be an error. In a 16-bit A/D application the least significant bit for a 10-V signal is 0.15 mV. This signal along the ground plane is a common-mode signal that must be rejected. This is another application for the forward referencing amplifier. In many low-level A/D converters, this amplifier is included in the input-sensing circuit. Care must be taken to connect the input signals to the points of interest in the circuit.

The signal source impedance affects both the gain and the CMRR. For amplifier sources the output impedance is inductive. For video signals the cable matching impedance must be considered. If the feedback resistors R_1 through R_4 are around 30 kΩ these effects will be minimum.

7.9 DRIVEN POWER STAGES (ISOLATED OUTPUT)

In single-ended designs it may be desirable to ground an output circuit without forming a ground loop. A useful circuit uses the output common

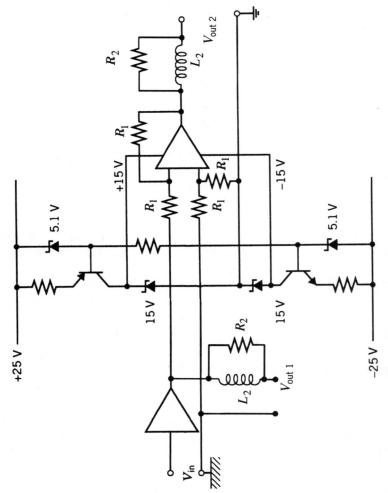

Figure 7.5 An output stage with common-mode rejection.

DRIVEN POWER STAGES (ISOLATED OUTPUT)

to drive the power supply. This power supply can be as simple as two zener diodes. A typical circuit is shown in Figure 7.5. The output stage is differential to reject the common-mode signal. Note that there is some ground current flow in the output common. For a short signal run and for high-level signals, this will usually be of little concern. Any errors will be small compared with the errors caused by a ground loop. This circuit technique can reject a few volts of ground potential difference. If large voltages are expected, then different techniques are required. If additional power supply voltage is available, then the zeners can be operated from two constant current sources. In either circuit, load currents must add or subtract from current in the zeners. The static zener current must be greater than the expected load current.

The parallel resistor and inductor in series with each output (R_2 and L_2) is suggested to keep capacitive loads from introducing circuit instability. An inductor of 10 μH and a resistor of 22 Ω is suggested for all outputs. The frequency response at 100 kHz for a 100-Ω load is only slightly modified.

8

CIRCUIT DESIGNS

8.1 INTRODUCTION

This chapter discusses analog circuit approaches that resolve many of the issues discussed in previous chapters. There are many ways to approach circuit design. Sometimes the obvious is apparent only after seeing the way another engineer has solved a problem. For many users this is all very academic. The important issue is: How does a design work in the real world? Somewhere between design and application there is a lot of information that may not get passed forward (backward). The user is better off knowing the issues that must be faced and the compromises that often must be taken. The conscientious designer thinks of many parameters often not mentioned in a specification. The user recognizes many shortcomings not considered by the designer. It is rare to have a designer with user experience or a user with design experience. Sharing problems and design experience helps to bridge the gap.

Specifications tell only a part of the story. Behind every specification there should be a test method. With out this method in mind, it is surprising how many assumptions are made by both the writer and the reader. Common-mode rejection is a good example. CMRR is often specified at one frequency for one line unbalance value and for one gain. The fact that an excitation supply produces a second CM signal is ignored. The high-frequency CMRR is usually not mentioned. It also turns out that CMRR is a function of gain setting. In fact a full CMRR evaluation would require perhaps 10 different specifications. The CMRR specification is picked as an example because it is at the heart of operating instrumentation between two grounds. CMRR is usually referred to the input (RTI). For example, if a 10-V common-mode signal causes an output error of 10 mV and the gain is 1000, the error referred to the input is 10 μV. The CMRR is thus 10^6 or 120 dB. A suggested wording for this specification is given in Section 8.3.

8.2 COMMON-MODE REJECTION—POSTMODULATION

When a signal is amplified in the input electrostatic environment (inside the input guard shield) and the output is observed in that same environment, the common-mode issue is not raised. If the signal is sent via radio to a remote point (telemetry), the CMRR would be infinite. The problem occurs when the coupling to the output environment occurs in the same enclosure.

It is practical to amplify a signal in the input environment and use a postmodulator/demodulator to "isolate" the signal. This system can use a switching frequency of 100 kHz and achieve a bandwidth from dc to 25 kHz. The carrier signal that results is transformer coupled. The postmodulator/demodulator may have a gain of one. The coupling across the transformer is usually not feedback-controlled, and this can result in linearity errors. Any carrier coupled back into the input can cause regeneration errors. These problems make this a difficult design problem.

The common-mode errors for this type of instrument are all generated at the input through some leakage reactance. This means that the CMRR referred to the input (RTI) is a constant and does not depend on gain. The common-mode voltage levels are determined by the breakdown voltages in the modulation/demodulation transformers. With care in layout, a short-term 5000-V common-mode signal can be accommodated. This may not be a steady voltage unless all dielectrics are rated for this stress, and this includes connectors. This postmodulator design does not require a balanced input stage, although this is highly desirable. The fact that there is gain in the input environment requires that there be a multishielded power transformer. All of the arguments given in Section 5.12 must be considered.

8.3 THE COMMON-MODE ATTENUATOR

The attenuator method of common-mode rejection is an important tool. The signal of interest is first amplified and then buffered. This buffering

THE COMMON-MODE ATTENUATOR

must provide a low source impedance to drive the following balanced attenuator. This buffered signal can be balanced or single-ended. The balanced attenuator then terminates on output common. Figure 8.1 shows this arrangement for both a balanced and unbalanced source. The attenuator reduces both the normal and common-mode signals. If the attenuator has a loss ratio of 30:1, a 300-V common-mode signal is reduced to 10 V. The gain following the attenuator is differential and can reject 10 V of common-mode signal. If the output amplifier has a gain of 30, the 300-V common-mode signal is rejected and the signal has the gain of the input stage. Obviously, more gain can follow this point.

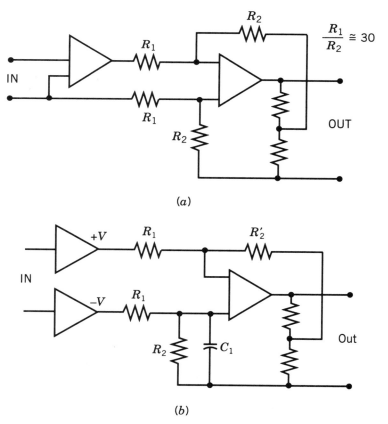

Figure 8.1 (*a*) Balanced attenuator with unbalanced source. (*b*) Balanced attenuator with balanced source.

An attenuation by a factor of 30 requires gain ahead of this point that exceeds 30. The reason relates to the signal-to-noise ratio. If the noise RTI (referred to input) is 2 μV and the input noise of the stage after the attenuator is 2 μV, the total noise RTI becomes $2\sqrt{2}$ μV. A gain of 100 ahead of the attenuator limits the RTI noise to 2.1 μV. Obviously, this added gain is not practical when the total gain requirement is only 30. There is an obvious penalty associated with attenuating signals in the signal path. The benefit is the ability to operate in the presence of high common-mode voltages.

The common-mode attenuator resistors must have a very low voltage coefficient of resistance. Typical metal film resistors are marginal. Several wirewound resistors in series can be acceptable. These resistors should be mounted clear of any other components. Special metal film resistors are available on the market that are well-balanced parasitically. These resistors can hold an excellent balance out to 100 kHz for voltages up to 300 V. In most applications an additional potentiometer is required to obtain a circuit balance to within one part in 10^6. A small reactive balance C_1 can be added to the attenuator to accommodate phase shift in the amplifier. This capacitor is shown in Figure 8.1b.

When a balanced attenuator is used for CMR, some of the common-mode error is generated at the attenuator and some by the following differential amplifier. This error is multiplied by the gain following the attenuator. The error referred to the input is the output error divided by the overall gain. As gains are changed the CMRR RTI changes. The CMRR RTI depends on the gain preceding the point of error. The user only needs to know that the error remains small when observed in the output for any gain setting. A simple and preferred specification would require that the output error be below 2 mV rms for a 300-V common-mode signal at 60 Hz for all gains of 1000 or less for a 1000-Ω line unbalance on either input lead for all gains. The CMRR for common-mode signals above 60 Hz should not fall off any faster than the ratio of frequencies. If the CMRR is 120 dB at 60 Hz, it should be at least 100 dB at 600 Hz.

8.4 HIGH-INPUT IMPEDANCE CIRCUITS

High-input impedances are required to provide a good CMRR when there is a line unbalance. These high-input impedances are provided by applying feedback to active circuits. FET gates have high-input impedances, but their RTI noise contribution is much higher than a junction transistor. Even

HIGH-INPUT IMPEDANCE CIRCUITS

with a FET, feedback must be used to limit the input capacitance. The best input circuit to use is the matched or dual transistor pair formed on one substrate. This matched pair is used in IC amplifiers, in IC instrument amplifiers, and in instrumentation amplifiers.

The basic feedback circuit for a high-input impedance single-ended input is shown in Figure 8.2a. The feedback causes the emitter to follow the input base signal. If the base moves 100 mV, so does the emitter within a very small error. The change in input current for the 100-mV change in input voltage is near zero. The ratio of voltage change to current change is the input impedance. The voltage gain of the circuit is the ratio of R_1 to R_2. This same basic circuit for a balanced signal source is shown in Figure 8.2b. The output signal is the voltage difference $V_{02} - V_{01}$. The gain determining feedback resistor is R_1 which goes between the two emitters. The gain is

$$G = 1 + \frac{2R_2}{R_1} \qquad (1)$$

Note that in this circuit the gain cannot be less than unity.

This circuit can be further enhanced by driving the associated power supplies (a string of zener diodes, for example) from a derived common-mode signal (see Section 6.8). The current for the input transistors is supplied from the output of the amplifiers through the feedback resistors R_2. This requires that the average voltage $V_{01} + V_{02}/2$ must be negative. The result is a balanced low-impedance output signal with a dc common-mode offset voltage. This offset voltage is removed by the following differential stage. A differential input impedance in excess of 1000 MΩ is practical with this circuit.

The current in the input transistors should be about 200 μA. The collector resistors can be about 20 kΩ and provide a bandwidth greater than 100 kHz. With this low collector current the base current that must be provided is about 2 or 3 nA. This current can be supplied from the driven power supplies through 100-MΩ resistors. Zeroing this base current is a factory adjustment in most instruments. When the adjustment is correctly made, then very little base current flows in the signal source impedance. A current of 1 nA in 1000 Ω is a 1-μV error. This current unfortunately is temperature-dependent, and a true balance can only be achieved at one temperature. For this reason it is desirable to keep the initial base current as low as practical. Transistor pairs with high β are obviously preferred because this limits the initial base current.

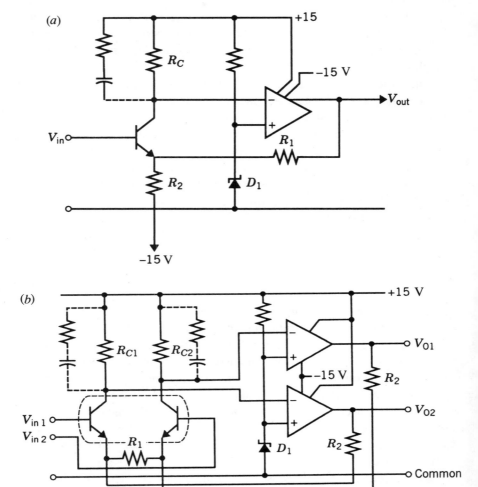

Figure 8.2 (a) High-input impedance circuit. (b) Balanced high-input impedance circuit.

8.5 BALANCED EXCITATION SUPPLIES

The arguments for a balanced excitation supply were given in Section 7.3. In review, the centertap allows the user to omit two bridge resistors when one or two active arms are used. The tap provides one of the input connections. A small series resistor in series with the signal lead allows the user to use the excitation supply to inject balancing signals, offset signals, or

BALANCED EXCITATION SUPPLIES

calibration signals. A precision (stable) zener voltage can be supplied that references the centertap. The presence of a centertap makes it easier to calibrate or substitution signals. A simple circuit is shown in Figure 8.3. Other methods of signal calibration can still be used.

A balanced supply allows the use of a full-wave centertapped transformer. This is a very efficient use of the transformer. Note that the entire secondary coil draws current each half-cycle. The centertap allows both a plus and minus supply. If the positive supply is regulated and adjusted, the minus supply can simply track an equal and opposite voltage.

Design problems exist when the balanced supply is a constant current source. A preferred method is to make the positive supply constant current and allow the negative supply to track the resulting output voltage as a voltage source. Two current sources cannot be placed in series. A voltage source in series with a current source appears to the load as a single current source.

A good circuit should have current limiting to protect transducers. Protection is automatically provided when the excitation source is used in a

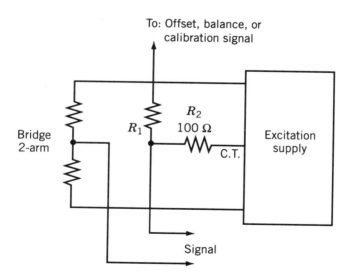

Current in R_2 provides an offset for the signal

Note: A precision 10 V reference voltage can be provided that uses the centertap as common

Figure 8.3 The excitation supply used to provide offset or calibration.

current mode. Excitation circuits can be designed to serve as either a voltage or current source. The voltage and current limits can be set externally by a digital-to-analog converter (DAC). In this type of design the load value can determine the mode of operation. Assume a 20-V supply with 100-mA current limiting. If the load is 1000 Ω, the voltage regulates to 20 V and current limiting is not invoked. If the load is 100 Ω the current limit is reached at 10 volts and the supply functions as a current source. Thus by providing current and voltage settings the excitation supply automatically finds its own mode of operation.

Stabilizing a dual mode circuit, when the load resistor matches both the current and voltage settings, is nearly impossible. A good approach is to place a small amount of hysteresis in the circuit so that the supply always switches between modes. This guarantees that there is no area of instability in the design. The user can always use voltage and current settings that avoid this load conflict. In the voltage mode the current limit should be set above the expected current. In the current mode the voltage should always be set above the expected load voltage.

The benefits to a dual mode centertapped excitation supply are apparent when an instrument is computer-controlled. The instrument can be calibrated, the bridge balanced, and the excitation modes set with DACs that are referenced to the centertap. The instrument can be digitally zeroed or offset even when a bridge is not involved.

8.6 CHARGE AMPLIFIERS

Charge amplifiers are used to amplify signals from piezoelectric transducers. These transducers are used in vibration studies where acceleration must be measured. The transducer is a crystal that looks electrically like a capacitor. When forces are applied to two opposite crystal surfaces a voltage appears. This voltage and the crystal capacitance imply a stored charge. This developed charge is converted to a voltage by a charge amplifier.

The voltage on the transducer could be amplified by a high-impedance amplifier. The problem here is that cable capacitance attenuates the signal. In a charge amplifier the cable capacitance increases the noise RTI, but it does not change the signal level.

The charge-to-voltage conversion takes place in an operational feedback circuit. The equivalent resistive feedback circuit is shown in Figure 8.4. The signal at the summing point ① is near zero. This point is called a virtual ground because the signal at this point is near zero. The input current is

CHARGE AMPLIFIERS

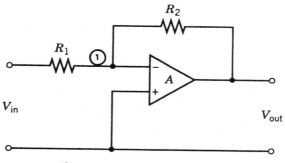

Figure 8.4 Operational feedback.

V_{in}/R_1. This current flows through R_1 and R_2 in series. The output voltage is the input current times R_2. The voltage gain is

$$G = -\frac{R_2}{R_1} \tag{2}$$

When feedback capacitors are used instead of resistors the voltage gain is given by the ratio of reactances or

$$G = -\frac{C_1}{C_2} \tag{3}$$

This is an ac gain as the reactances go to infinity at dc. A high-value resistor shunting C_2 can be used to limit the low-frequency response. This resistor also provides a path for any input base current and keeps the circuit from overloading at dc. The charge converter is shown in Figure 8.5.

Capacitor C_1 can be considered the transducer. Charge generated on this capacitor is transferred to C_2. In a practical circuit the transducer (capacitor) is at the end of an input cable. Charge generated by the transducer is transferred to the feedback capacitor. This results in an output voltage. The summing point ① has essentially no signal voltage. The summing point extends from the instrument to the transducer. The output voltage is

$$V_0 = \frac{Q_1}{C_2} \tag{4}$$

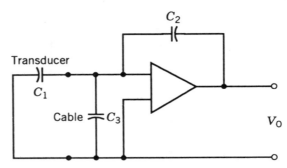

Figure 8.5 A simple charge converter circuit.

The gain of this charge converter circuit is controlled by C_2. The smallest practical value is about 200 pF. This circuit must be very carefully shielded or there will be gain errors. As an example of the problem, consider a gain of 100 after the charge converter. A capacitance to this output of 0.02 pF represents a 1% gain error. A power supply ripple of 1 mV can be capacitively coupled to the input and seen as interference. It is easier to double regulate the power supply than to attempt any added shielding.

A low-noise input cable must be used for charge amplifier applications (see Section 6.5). This cable capacitance decreases the signal-to-noise ratio. The noise RTI is given in terms of a base number plus a figure related to the cable capacitance. This cable is usually single conductor shielded.

The charge converter circuit is single-ended. There have been attempts at providing balanced signal transducers, but this is a difficult manufacturing problem. A balanced signal source would require a balanced charge converter circuit which is not available. These attempts at balancing charge input circuits results from the need to ground the input transducer. A single-ended charge amplifier cannot be multiply grounded or a ground loop results. The proper solution is to make the entire charge amplifier differential. This does not require balanced inputs, but it does require an understanding of how to reject common-mode voltage.

The transducer itself is a mechanical device. It has mass and at high frequencies it may have its own mechanical degrees of freedom. Transducers must be selected so that they do not modify the device they are testing. When a transducer is mounted on an insulator the insulation adds a degree of freedom to the mechanical system. It is preferred to ground the transducer and avoid this layer of insulation.

A charge converter with a postattenuator and postdifferential stage can be used to reject the ground potential difference. The charge converter requires its own power supply, and this requires a multishielded transformer. The input shielding requirements are best met using special low-noise cable.

8.7 CHARGE AMPLIFIERS AND CALIBRATION

The ideal calibration system passes a signal through all parts of the system. Hitting the system with a calibrated hammer may seem crude, but it does check every component in the vibration measurement. The injection of electrical signals for calibration has its problems. A separate capacitor can inject a calibrated charge into the input. Unfortunately, this mode of calibration cannot detect if the cable or transducer are disconnected.

The injection of charge must be a function of the gain setting because the gains can vary over a 1000:1 range. One approach is to normalize the gain for calibration, and the second approach is to attenuate the calibration signal as a function of gain setting. The first approach does not check the instrument at the gain of interest, and the second approach raises the cost and introduces a reliability issue.

A calibration scheme that includes the input cable and transducer involves a voltage inserted in series with the input shield. The injected input charge that results is a function of the cable and transducer capacitance. With this calibration method an open cable or open transducer can be detected. Because of undefined cable lengths the calibration is not accurate, but it is at least repeatable.

A reliable calibration procedure checks for accuracy and for the presence of the transducer. This may take two separate procedures. The hammer procedure is simple and does show that the transducer and cable are in place. An injected charge shows that the gain is correct. With today's computer-controlled signal generators a specific signal level for each instrument is practical.

8.8 CALIBRATION GUARDING

To be useful a calibration signal (signal substitution) must involve the input to each instrument. Leads that thread through every instrument and couple to each input can be a source of interference and cross-coupling. A calibration signal is referenced to a ground that is different than the input or output ground. It is costly to provide an individual transformer for each instrument to isolate this signal. It is best to bring a substitution signal into the output environment. A unity gain forward referencing amplifier (see Section 7.8) can reference this signal to output common. This signal is then forward-referenced to the input common. This way a contaminating ground does not enter the input environment. The common-mode levels during calibration may only be a few volts so that the isolating electronics can be kept simple. This method of isolation is less expensive than the use of relays. When not in use the substitution signal should be made a zero-impedance source with zero signal.

In some cases, sufficient isolation can be provided by using guarded relay contacts. During calibration the three input leads are switched from their normal positions to a signal bus. The new guard potential is the ground of the substitution signal. During normal operations the open contacts on the relay are connected to the local guard potential. This guarding takes extra relay contacts. This guarded configuration is shown in Figure 8.6. With this arrangement the leakage capacitance out of the input guard can be held to a few tenths of a picofarad.

8.9 FILTERING ANALOG SIGNALS

Noise that interferes with a signal can be generated in the transducer or coupled from the environment. If the interference is inside the band of interest, there is no simple filtering algorithm that can remove the noise.

Filters are often placed in the signal path to attenuate unwanted interference or unwanted signal. In dc instrumentation the filter is usually placed

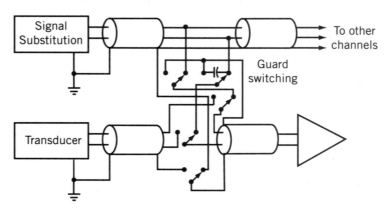

Figure 8.6 Guarded relay contacts.

after the last block of gain. This is necessary to avoid a drift component from the active filter circuits. Filters with drift levels of 1 mV RTO are acceptable. If the filters are placed ahead of the last gain block, the drift error from the filter must be correspondingly lower.

Filtering requires "headroom." The signal must not overload the output stage or the result is nonsense. If necessary, the overall instrument gain can be reduced to avoid overload. Overload conditions should always be considered before filtered data are accepted.

Filtering can occur after data have been digitized. These digital filters introduce no drift. Care must be taken to avoid aliasing errors. This error can result when there is signal content above one-half the sampling frequency. An example of aliasing error occurs in a motion picture when the wagon wheels appear to go backwards. Sampling frequencies above 100 kHz are practical. Storing filtered sampled data requires far less memory than storing the raw sampled data.

State variable active filters lend themselves to computer control of pole and zero locations. This type of filter requires more active circuits than the conventional two- or three-pole circuits. The active circuits are either integrators, summers, or multipliers. These circuits can be complex enough to warrant placing them in another instrument package.

Often it is necessary to maximize bandwidth and limit sampling rates. This places a squeeze on the filter design. To resolve this problem, eight-pole elliptic filters have been used. This is an expensive solution to data acquisition. As sampling frequencies go higher and digital storage becomes cheaper the filtering problem becomes less severe.

Filters with very steep cutoff characteristics have bad transient responses. Step function overshoots in excess of 40% are not uncommon. If the data are transient in character, then this overshoot may be unacceptable. Also note that pulse-type noise can also cause transient overshoot. Bessel filters (linear-phase filters) have no overshoot, but they have a very rounded amplitude response shoulder. Extra poles do not alter this round shoulder effect. Nonstandard filters can be designed to optimize both step function overshoot and out-of-band roll-off. These filters have a known frequency response, and the data are compatible with any form of analysis.

8.10 THE AC AMPLIFIER

Coupled ac amplifiers are used in audio designs, charge amplifier designs, telephone circuits, and so forth. The fact that a dc response is not needed implies that signal transformers can be used. In telephony, transformers are used to convert balanced signals to unbalanced signals, and vice versa. Gain is only provided to accommodate any line losses. In charge amplifiers, transformers are not used because of the very low frequencies that are involved. Because of cost the conversion of signals from balanced to unbalanced mode is approached actively where possible. In audio systems the active approach is used because the common-mode levels are apt to be small. The active approach can only be used when the common-mode levels are under control.

When transformers are used there are two ways to approach the design. The transformer can be outside of any feedback loop or inside. At the input where power levels are low, feedback is usually not used. In output circuits the merits of placing the transformer inside the loop involves a reduction of distortion, improvement of frequency response, and control of output impedance. Stabilizing a feedback circuit with a transformer in the gain path can be troublesome. There can only be one dominant pole of transmission at low frequencies. In other words the low-frequency response must roll off at 20 dB per octave until the loop gain is unity. If this is not done, the system will be unstable. The dominant pole can be provided by a coupling capacitor or the transformer magnetizing inductance.

In ac systems where several coupling capacitors are used, the recovery from overload can pose problems. If there is gain following a coupling capacitor, the recovery time is increased. It is possible to design ac systems with a transient response having one long undershoot. An undershoot is necessary because the average output signal must be zero. Systems should always be tested to see that the recovery process is acceptable.

Carrier systems are usually ac-coupled. When the carrier is turned on or off, a transient results. This transient can be troublesome for some loads. In large shaker designs it is necessary to add circuitry to bring signals up slowly and to limit output voltage swing upon power shutdown. If this is not done, then the shaker might be damaged as it bangs against the stops.

8.11 SUPPRESSION CIRCUITS

Relay or solenoid contacts that arc can radiate energy throughout a facility. This radiation can be disruptive to computer circuits and can easily cause interference for video signals. Nearby circuits can actually suffer damage. The reason for the arcing is not always appreciated.

SUPPRESSION CIRCUITS

When a contact opens on an inductive load carrying current, the current must continue to flow. The reason is simply that there is energy stored in the inductance. This energy cannot be made zero in zero time because this takes infinite power. The current starts to flow in the parasitic capacitance of the coil. In the case of a solenoid or relay, this capacitance is on the order of a few hundred picofarads. This capacitance and the device inductance form a resonant circuit where energy transfers back and forth. The resonant frequency is rarely above 10 kHz. Energy normally would transfer to the capacitance in the first quarter cycle of the resonant frequency. When the current is zero the voltage across the capacitance is maximum. Consider a relay with 1-H inductance and 200-pF capacitance. If the relay current is 100 mA, the energy stored in the inductance is $\frac{1}{2}LI^2$, or 0.005 joules. The energy stored on the capacitance is $\frac{1}{2}CV^2$. This voltage calculates to be 7000 V. The voltage rises to its maximum in 25 μsec, or one-quarter cycle, at 10 kHz. In this time the relay contact has barely opened. This high voltage across a small distance breaks down the air and an arc results.

The arc radiates energy over a wide spectrum that extends well beyond a gigahertz. There are several ways to limit the voltage so that this arcing does not occur. The simplest way is to place a reverse diode across the coil. When the contact opens, the current in the relay or solenoid flows in the diode. The voltage is limited to the forward drop in the diode. The current decays exponentially as determined by the L/R ratio, where R is the resistance of the coil. Typically a relay with a shunting diode may stay operated for a tenth of a second or longer. In telephony the dropout time of a relay is critical and diode suppression is not usually acceptable. The relay dropout time can be shortened by adding a resistor in series with the diode. This resistor can be adjusted to limit the voltage so that arcing does not occur.

A second way to limit the voltage across the contacts is by adding capacitance across the relay or solenoid coil. This capacitance stores the energy at a reduced voltage, and it takes longer for the voltage to rise. The capacitance can be adjusted until there is no arcing. In many installations it is impractical to add suppression circuitry to relays or solenoids, and the equipment must be designed to accommodate this interference. This places a burden on both equipment and systems designers. Interference coupled to cables can be very disruptive. This is an area not easily controlled by the equipment designer.

Placing solenoids or relays inside metal enclosures is not always a solution. The reason is that radiation couples to every lead that enters or leaves

8.12 SWITCHING REGULATORS

The key to switching regulators is the storage of energy in a magnetic field. This stored energy can be transferred by transformer action to an isolated load. A switch closure allows the energy to build up in a magnetic field. When the switch opens, the energy in this magnetic field is transferred into a capacitor. Load resistors then take energy from the capacitor. The longer the switch is closed, the more energy is stored. The number of switching cycles per second and the energy store per cycle determine the power level of the regulator. The control of the switch is available as a standard integrated circuit.

Magnetic field energy is stored in a gap. Ferrite cup cores with a gap are commercially available. A typical circuit is shown in Figure 8.7. When the switch closes current starts to flow in the inductance. When the switch opens, the current must continue to flow and somehow return to zero. The only way the current can begin to diminish is for the voltage across the

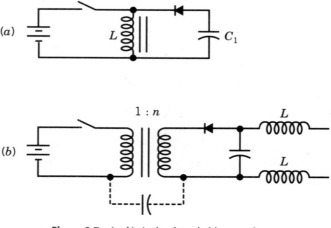

Figure 8.7 (a, b) A simple switching regulator.

SWITCHING REGULATORS

coil to reverse polarity. This is the direct result of applying Faraday's law to an inductance. This reversal of polarity allows the current to flow in the diode D_1. This current can now store charge in the capacitor C_1. Thus the energy stored initially in the magnetic field is transferred to the capacitor as an electric field.

When the switch opens, the voltage across the switch rises very rapidly. At the moment of switch opening, the voltage at the inductance reverses polarity. There is no loss of generality if a second coil is used to couple the stored energy to a capacitor. This circuit is shown in Figure 8.7b. The turns ratio allows the transfer of energy to occur at a different impedance level. If the secondary turns are reduced, the circuit allows for a high-current power supply. If the turns are increased, the secondary voltage can be higher than the primary voltage. By adjusting the duty cycle the voltage on the secondary can be regulated.

The secondary circuit is isolated from the primary circuit by transformer action. The capacitances from the primary to the secondary allow parasitic currents to circulate. Since the primary voltage contains step functions, the transformer capacitances can couple interference into secondary circuits. The interference is in the form of spikes at each voltage transition.

Filters placed on the secondary circuits must limit the flow of interference current on both leads. Note that the secondary circuit is usually grounded on one side. The degree of filtering depends on the application and on code regulations that apply. Any interference current that flows past the filter must return to the primary via the grounding system. It is desirable to return this current directly to the primary circuit rather than have it circulate throughout a facility to return. It is obviously desirable to provide a product that limits the currents that circulate in external conductors. Not all products are equal in this respect.

The transformers used in switching supplies operate at frequencies above 50 kHz. The number of primary turns in the transformer is usually small. This class of transformer does not lend itself to internal shielding because the added capacitances limit high-frequency operation. The only way to control the flow of parasitic currents is through symmetry in the transformer. The secondary coil must capacitively couple to both plus and minus transitions at the same time, thus greatly reducing the interference current that must be filtered. This symmetry is available if the primary and secondary coils are both centertapped. Every turn on the secondary must parasitically couple equally to both halves of the primary. This symmetry requires care and attention during construction of the transformer.

8.13 OFF-LINE SWITCHERS

The isolation provided by switchers makes it possible to pull energy directly from the power line rather than from some intermediate filtered dc source. This technique does not require a transformer. A full-wave rectifier on the power line provides a dc source that can be used to operate a switching power supply. If the power is supplied from a three-phase source, then power is available over the entire cycle.

The rectifier system may or not have associated storage capacitors. If storage is not used, then energy must be taken directly from the power line in small gulps. These pulses of current place small "glitches" on the power line that are sensed by every nearby user. Line filters to the equipment ground conductor and across the line can limit any power-line spikes. If reactive energy storage is used (filter capacitors), then the energy can come from the storage rather than from the line. When the diodes of the rectifier are conducting, the energy is again taken from the line. This results because conducting rectifiers are a low-impedance path back to the power lines. Line filters, if used, must be able to supply the high currents required by the switcher. Another technique that works is to avoid pulling power from the reactive storage when the diodes are connected to the line.

All of these techniques imply that there are periods in the power cycle when energy is not available or when it should not be taken from the line. During these intervals, energy must be taken from local storage so that the switching supply can still function. These techniques are in the domain of the power supply designer. The user should verify that the interference caused by an off-line switcher is limited and does not create a problem. The user should also investigate the impact of large filter currents flowing in equipment grounds.

8.14 MOUNTING SWITCHING TRANSISTORS OR FETS

The collector or drain connection to a switching transistor or FET is the case of the component. This connection provides a thermal path for heat flow. The user must make sure that the component is used with a heat sink to limit the temperature rise. In high-power applications this may mean that the transistor is mounted directly on the chassis of the hardware. In most circuits the collector or drain cannot be grounded and it must be insulated from its mounting point. It is standard practice to use insulating washers and a gel to aid in heat conduction. The gel is not as good as a solid ohmic contact.

The capacitance from the case to the mounting ground can be 100 pF. If the collector or drain voltage changes at 100 V/μsec, the current in the capacitance is 10 mA. This current must return to the power supply via an equipment ground path. This path can be inductive (loop area), and radiation and interference can result. There are ways this current can be reduced and the loop areas limited.

A commercial product called SIL-PAD can be used as the component insulator. This pad is a thin sheet of metal sandwiched between two insulators. The metal portion has a tab that allows it to be connected to the circuit common. The SIL-PAD forms a guard shield that shunts most of the parasitic current back to the power common without using the equipment ground. The leakage capacitance around the SIL-PAD might be 2 pF. If the heat sink is used for the power supply ground connection, the return loop area can be easily controlled. A heat conducting gel should still be used. (See Figures 8.8 and 8.9.)

8.15 PARALLEL ACTIVE COMPONENTS

It is often necessary in design to use parallel active elements. This may occur in power supply series regulators or in output drivers. When two or more elements are paralleled, the resulting circuits are susceptible to high-frequency instability. Suppression resistors must always be added in series with these connections. For example, when transistors are paralleled, resistors should be placed in series with the emitters and bases. The base resistors may be as low as 20 Ω, and the emitter resistors may be as low as a 0.5 Ω. The emitter resistors also serve to balance the current between transistors. When instabilities occur, they can be at frequencies well above 100 MHz and go unnoticed until the component self-destructs. Better grounding or shielding has very little to do with this form of instability.

8.16 THE MEDICAL PROBLEM

The body develops many electrical signals that are often monitored. The standard signals are EKG (signals from the heart) EEG (signals from the brain), and electromyogram (signals to muscles). To monitor these signals, conductors must be placed on the skin. Subcutaneous probes are superior

CIRCUIT DESIGNS

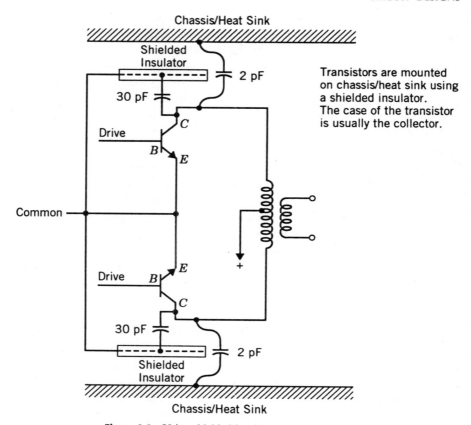

Figure 8.8 Using shielded insulators to mount transistors.

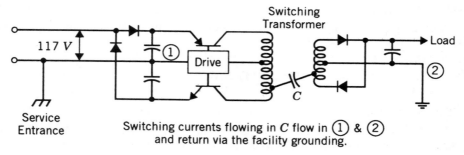

Figure 8.9 Coupling across a switching transformer.

THE MEDICAL PROBLEM

but not commonly used. Techniques for obtaining stable low-resistance contacts to the skin have improved significantly over the years.

No one point on the body is a best ground. The signals that are generated in the body flow throughout the body. Muscle signals mix with heart signals and are a form of interference. Specific reference points are selected to optimize those potential differences that provide the most information on body function.

External fields cause currents to flow in the body. These currents add interference to the signal of interest. The use of differential high-impedance amplifiers cannot eliminate this differential interference. Measurements are best made in an environment of low E field. The signals can be filtered to eliminate out-of-band differential components of power coupling. The common-mode signals are rejected by a suitable differential input. The shields on the input conductors are connected to one point on the body which is the guard potential.

Most of the instrumentation involves ac amplifiers. This eliminates the dc contact potentials that can be troublesome. The drawbacks are phase shift and low-frequency transient response. If the low-frequency character is controlled, the data can provide the signature information required by the observer. With dc restore circuits it is possible to have dc transient behavior and avoid the issues of contact potential.

Grounding the body to an earth point is of questionable value. This invites more parasitic current to flow in the body. In some cases the interference may be reduced, but there is no guarantee. The issues of ground current flow generated by the power transformer all pertain. If the output is isolated (graph plots on paper), then the instrumentation is simpler to implement. This material is covered in detail in Chapter 5.

The leakage currents that flow in the body as a result of electrical connections must be kept quite small. The regulating codes that relate to this problem limit the amount of equipment ground current that is permitted. This current flows in filter capacitors going between the power line and the equipment ground. Depending on the equipment involved, this current may have to be less than 10 μA. If the equipment ground path is broken, the current path would be through the patient. Without this limitation the resulting body currents can become lethal. In a hospital environment the threat to life is obviously greater. This is one of the reasons why floating power is sometimes requested. It must be remembered that floating power still has a capacitance through transformer windings back through voltages to the service entrance earth and current can still flow.

8.17 THE MAD COW PROBLEM

Milking machines can pose a problem to farmers. If there is a current path from the milker through the cow, the result is an uncooperative cow. The problem can be likened to the millivolts generated in the human mouth. When a piece of aluminum foil forms a battery with a metal filling, the effect is very painful. Tiny voltages are interpreted as pain and there is significant discomfort.

Wet soil conditions under the cow provides a path to the power system grounding. If the milker is not clean, then the surface provides a conductive path through the udder to earth. Very small potential differences can cause difficulties. This problem can be resolved by careful design and by maintaining a clean environment.

9

UTILITY POWER

9.1 INTRODUCTION

Utility power couples into every piece of electronic equipment. It also supplies lighting, motor controllers, elevators, computers, monitors, appliances, and air conditioning. It is obvious that many interference problems can arise from this complex of loads. Fortunately, in large facilities the motor and lighting loads are usually supplied from separate distribution transformers. This separation in power distribution provides a separate neutral (grounded) conductor and a separate equipment grounding system. The equipment grounding conductors can ground (earth) at many points, and this allows some interference currents to circulate in the facility. The fields associated with this circulation can impact the performance of sensitive equipment.

Individual analog instruments provide the user with an undefined grounding system. When an output common is connected to a terminating device, this defines the output reference potential. Another way of saying this is that the amplifier is "grounded" by the output connection. If the input is grounded at the transducer, then the reference potentials for the instrument are defined by its connections. If the instrumentation is designed correctly, then very little current will flow between the input and output grounding through the instrument. If current does flow, it flows in shields and common-mode attenuators but not in input signal conductors.

Digital devices and single-ended devices that are not differential in character can allow current to flow from input to output on signal leads. If the signal levels are high and the leads lengths are short, then common-impedance coupling can be tolerated. This is frequently the situation on a test bench. When the same equipment is distributed in a facility, then the extended lead lengths can allow increased coupling and the interference can be objectionable. This is a common problem and it needs special attention. This chapter treats the ground plane as one possible solution.

Power-line filters are often improperly mounted. This usually makes the filters ineffective at high frequencies. This chapter deals with this problem. Special distribution transformers (isolation transformers) can be used to limit many interference processes. This isolation transformer is very different than the shielded transformers discussed in Sections 5.6 and 5.7. Filters, isolation transformers, and ground planes, when properly applied, can be used to reduce interference coupling. These topics are covered in this chapter.

9.2 THE GROUND PLANE

An ideal ground plane is a thin layer of metal associated with an area of interest. The use of the term *plane* is misleading because a bent or curved surface can be very effective. Ideally, equipment is bonded to this ground plane and all cables are routed near the ground plane. The ohms per square on a sheet of metal are microhms (see Section 4.2). If stray currents flow across this surface, the potential differences are at most millivolts. The only electromagnetic fields that can propagate over this surface are fields with a vertical E component. This means that cables routed on the ground plane will not couple to this field because there is a minimum loop area (see Section 4.11).

Many types of ground plane have been constructed. One approach is to use the rebars buried in the concrete. There are many problems with this technique. First, it is difficult to bond equipment to the rebars where required. Second, there is no guarantee that the rebars form a welded grid. Simply having an earthed conducting grid under the equipment is not a solution to anything.

Large electronic systems are often built on a raised floor. The floor can serve as a ground plane. This type of floor consists of metal stringers on 2-ft centers. The stringers are bolted together at each intersection. To be as effective as a thin sheet of metal, the resistance at each bond must be in microhms. This requires plated surfaces and the use of pressure washers.

SINGLE-POINT GROUNDING OF GROUND PLANES

The floor tiles are made slightly conductive (10^7 Ω per square) and they must make connection to the stringers. This treatment is needed to bleed off any charge accumulated on a person walking across the room. These conductive tiles help in avoiding problems with electrostatic discharge (ESD). The stringer system rests on stanchions. This forms a raised floor that provides a plenum chamber under the equipment that can be used for moving refrigerated air. Designers use this chamber to route cables and distribute power to the racks.

Cables that are routed on the concrete floor allow large loop areas between the floor and the cable. Common-mode coupling is proportional to this area, and obviously this is not ideal (see Section 4.11). It would be preferable to route cables along the stringers, but this is difficult to implement.

In a practical approach the ground plane would be placed on the floor under the plenum chamber. The plane could be strips of copper or steel bonded at each intersection. This ground plane would be bonded to each rack using a wide strip of metal. This extends the ground plane from the floor to the walls of each rack. Cables leaving the rack should follow this extension. This approach limits common-mode coupling to each cable. This approach also reduces the performance requirements of the raised floor, which need only be slightly conductive to limit ESD.

A ground ring is often placed along the perimeter of the room. All metallic conduit entering the plenum chamber is bonded to this ring. The ring, in turn, bonds to the floor and to the building steel at many points. If lightning were to enter this complex of conductors, there would be a very small voltage gradient and no equipment would be damaged.

Floor tiles present a unique problem when they are removed. If they are rubbed during removal, a charge can build up on the surface. The voltage on the tile increases as it is moved away from the stringers. If the voltage is high enough, there can be an arc from the tile frame to the stringers. This arc is an ESD that can couple to open cables that lie on the floor below. An example might be ribbon cable carrying logic. Placing the cables in a covered tray would help eliminate this problem. If this problem occurs, it is wise to check the grounding of the tile and also check that the tile is conductive.

9.3 SINGLE-POINT GROUNDING OF GROUND PLANES

Single-point grounding the ground plane violates the intent of the ground plane. This approach is sometimes taken in an attempt to keep stray facility currents from using the ground plane as a current path. Unfortunately, at high frequencies there is no such thing as a single-point ground. The capacitances to earth and to other conductors provide many parallel paths.

The problem becomes obvious at power frequencies when the equipment grounding conductors are considered. These conductors must connect the frame of every piece of equipment to the service entrance ground. This is required by code for fault protection. If every piece of equipment bonds to the ground plane and also to the service entrance ground, then a single-point ground cannot exist.

In some facilities the equipment grounding conductors are returned over a separate set of conductors to a central earth point (a star connection to a Mecca). This is another attempt to keep filtered current away from the ground plane. This type of connection violates the intent of the NEC and is probably ineffective. Power filters are usually mounted on a chassis which is rack-mounted. This chassis is also connected to the equipment grounding conductor. The result is that filter currents still flows in the ground plane through the rack as well as in the equipment grounding conductor. This star connection of conductors is a much higher impedance return path than the ground plane. This means that the filter current path is not controlled by rerouting the green wires.

Any concentration of current implies a higher inductance, and for a large current pulse this means a high voltage drop. The idea behind single-point grounding would be to limit ground current by limiting the number of connections to the building steel. Some designers might suggest adding an inductor in this path to further limit current flow. This practice is dangerous. If there is a lightning pulse, large potential differences can result and the inductance along with some equipment may not survive. Furthermore, if the ground plane is single-point grounded, this forces the fault path to be through this connection. Now both the star connection and the single-point ground are high impedances (large loops), and the safety of the facility is in jeopardy.

9.4 RACKS AS A GROUND PLANE

It is practical to bond racks together to form a ground plane. The hardware used should form a bond that will not corrode or come loose over time. A sheet of metal on the floor of the racks can be used to improve the ground plane between racks. If cables are routed on the floor, then this ground plane addition will be used.

9.5 EXTENSIONS OF A GROUND PLANE

It is often necessary to extend a ground plane to a nearby room. This cannot be handled by a single conductor no matter what the diameter. The best way to connect two ground planes together is to make connections across the entire boundary. The connection might be made with #10 AWG conductors on 6-in. centers. This technique maintains the mechanical integrity of the wall that must be breached.

A ground plane can be extended along a wire tray or metal gutter. Sections of the tray are usually connected together by sheet metal screws, and this is not an acceptable bond. The preferred bond involves welding or soldering. The most important bond is at the ground plane. An alternative method is to line the tray with a sheet of metal. Any metal will suffice provided that the installation is robust enough not to rip or tear. This added metal must be bonded to the ground planes on both ends of its run.

The impedance of a ground plane that is not square can be calculated by dividing the area into smaller squares. A tray that is 2 ft wide and 30 ft long has an impedance of 15 squares in series. If one square is 1 mΩ per square, then the tray has an impedance of 15 mΩ. It is easy to see why a wide tray may be acceptable while a single large conductor is not. A single conductor looks like an inductance, while a tray appears more like a resistance.

If it is necessary to connect ground planes located on separate floors of a building, then the ground planes must extend along one wall. It is easier to install a ground plane in a wall during construction than to install it later. Sheet metal, if properly bonded, can be used for the wall connection.

It is usually impractical to extend ground planes between buildings. Single connections are too inductive to be practical. If there is a lot of digital transmission between buildings, then fiber optics is a preferred solution. A break should be provided in any steel wires used to support the fiber-optics cable. This reduces the risk of a lightning pulse entering either facility over the optics link.

9.6 THE EARTH AS A GROUND PLANE

The resistivity of the earth varies depending on soil conditions and moisture. On a volcanic peak or on a granite outcropping, a good earth connection is impossible. Large volumes of earth have a low resistance, but a low ohms-per-square value is rarely available on the earth's surface. For this reason, radio transmitting antennas often rely on a metal grid placed on the earth under the antenna (counterpoise). This grid makes it possible to have uniform transmission independent of soil conditions. In the desert the sand is often an insulator and a reliable earth connection is often not practical.

For electromagnetic transmission the earth is a reflector. It is convenient to assume that the reflection surface is located at one skin depth. This depth varies with resistivity and frequency (see Section 4.4). The conductivity of sea water is low, and high-frequency radiation is rapidly attenuated. This is the reason why VLF transmitters are needed for underwater communications.

When an earth connection is not available, the issue of power safety must be reconsidered. If all of the conductors in a facility are bonded together to form a grounding electrode system, the facility is safe and an earth connection is unnecessary. This is the same issue as power on an aircraft. An earth connection on an aircraft would be a severe handicap. Obviously a floating system can be made safe without an earth connection.

The degree of bonding in a facility depends on the magnitude of the fault currents that are involved. In a generating station, fault currents in excess of 10,000 A might be expected. This current flowing in 10 mΩ is a voltage of 100 V. If this voltage exists between a metal floor and a rack, it is easy so see how this is a safety hazard. It is obvious that bonding for safety is a function of the current levels that might be encountered.

9.7 ISOLATED GROUNDS

An isolated ground as defined by the NEC is not a floating ground. Equipment grounding conductors (green wires) are normally connected to the conduit and receptacle as well as being connected to a green wire that eventually connects to neutral (grounded conductor) at the service entrance. In the isolated case the equipment grounding conductor is not connected to the local conduit or receptacle but is, instead, routed separately back to the service entrance or an intermediate panel. The idea here is to avoid sharing equipment grounding conductors with another device. The assumption is that an isolated ground will avoid common-impedance coupling that can impact the operation of the equipment.

Receptacles that "isolate" the ground must be colored orange or marked with a triangle or delta symbol. The corresponding plug need not be specially identified. The code prohibits modifying a receptacle to isolate the equipment grounding conductor.

The use of a separate equipment grounding conductor raises the impedance in this path. This reduces the effectivity of the filters in a piece of equipment. A line filter is normally connected to equipment ground and returns fundamental power current and noise to the service entrance via this conductor. This is the only path when equipment is not rack-mounted or when it is insulated from the rack. When the return paths have high impedance and several pieces of equipment are interconnected, some of the filtered currents will use the interconnection to return to the service entrance. If the signal paths are single-ended as in the case of a video signal, the signals are easily contaminated.

Computers that are interconnected should not be powered from isolated equipment ground receptacles. The difficulty arises when loads are switched on or off in the facility. At the moment of disconnect, there can be large inductive spikes on the power line. If equipment filtering is ineffective because of a high-impedance equipment grounding path, then these spikes can appear on the interconnection between computers. These high voltages can destroy input logic circuits.

9.8 SEPARATELY DERIVED POWER

Power generated locally, auxiliary power, solar power, UPS system (uninterrupted power system), or power from a facility transformer are defined as separately derived. The code requires that the neutral or grounded conductor for this power source be grounded (earthed) to the nearest point on the grounding electrode system for the facility. There can be only one grounding electrode system per facility, and this implies that all neutrals (grounded conductors) be connected together. It is preferred to bring all power into a facility at one point so the distance between these neutrals (grounded conductors) is kept to a minimum. This helps to limit potential differences between neutrals.

Each separately derived system is treated as a new service entrance. This means that the equipment grounds for the new feeders and branch circuits must return to the new source panel. The neutral or grounded conductor can be earthed only once, and this is at the new source panel. A separately derived system is the same as a new service except that there may not be a meter.

The beauty of a separately derived system is the new neutral carries currents for the loads on that system and no others. If the transformer is properly located, the neutral run can be very short. This means that the neutral conductor (grounded conductor) brings a relatively clean power to each piece of equipment. This is critical in electronic environments where the loads are very rich in harmonics.

9.9 POWER ISOLATION TRANSFORMERS

The words *isolation transformer* imply a solution to a problem. Just adding a transformer to a problem rarely works. The problem and the solution must match. An isolation transformer used in instrumentation usually has one or more electrostatic shields wrapped around internal coils. This transformer is inside of the equipment and is not considered a part of the facility. These shields are connected to various points within the instrument and can control the flow of unwanted current (see Section 5.6). When an isolation transformer is used to power a group of instruments, then some new issues arise. The transformer becomes a part of the facility, and the role of the shields is quite different.

The transformers and associated circuits that are described in this section can be single-phase or three-phase. The single-phase circuit is used as an example because it is simpler to draw. The single shield is shown in Figure 9.1. This transformer is ideally mounted on a ground plane. The single shield is internally connected to the transformer frame. When the transformer is mounted, the shield is connected. The primary and secondary conductors are carried in conduit. The conduit bonds to the transformer frame and is considered a part of the equipment grounding system.

Transformer action involves the difference signal between the power conductors. There can also be a common-mode signal (interference) on the primary transmission line. The shield in Figure 9.1 reflects any common-mode signals and stops them from coupling to the secondary circuits. Without the shield, capacitance couples common-mode signals directly from the primary to the secondary circuits. In a typical isolation transformer the leakage capacitance out of the shield is about 5 pF. With a shield, most of the common-mode current returns inside the primary conduit and never sees the secondary circuits.

Some of the common-mode current flows in coils of the primary. This current flow appears in the secondary by transformer action. To limit this effect, a second shield can be added to the transformer. This shield is

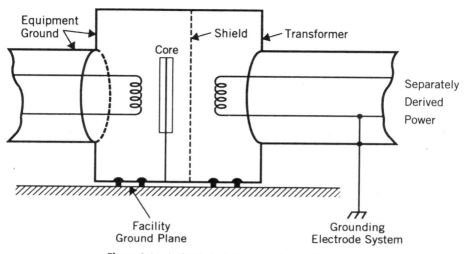

Figure 9.1 A simple isolation transformer (power).

internally connected to one side of the primary coil. Common-mode current now flows between the shields not in the coils of the primary. This added shield is shown in Figure 9.2.

If the secondary loads generate any common-mode signals, it is desirable to keep these signals from propagating through the transformer back to other systems. This path can be closed by adding a third shield that is again connected to one side of the secondary coil.

The power isolation transformer has internal shields that are not available to the user. If they are brought out separately, they should be connected at the transformer. Connections to remote grounds are too inductive. These connections make no electrical sense and they violate the isolation processes.

9.10 COMPUTER POWER CENTERS

The code permits the grounding of a separately derived power system on a computer floor (ground plane). The transformer is an isolation transformer. The power center includes the transformer, breakers, meters, and filters. The filters are discussed in the next section. Because it is a part of the facility, it may have to me mounted and wired by a qualified electrician. The concession the code makes is that it allows the computer floor to be

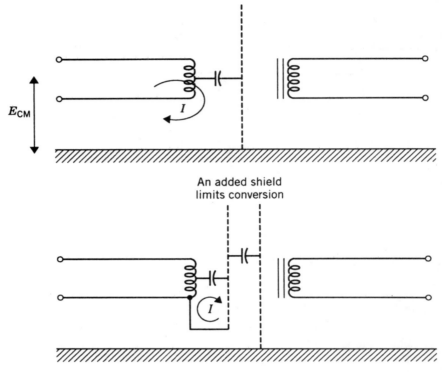

Figure 9.2 A two-shield isolation transformer (power).

considered a a suitable grounding conductor. Normally this conductor would have to be a qualified earthed conductor at the service entrance or a part of building steel.

The primary and secondary circuits should not share the same panel, conduit, tray, or outlet box. If this is done, the isolation provided by the power center is violated.

9.11 POWER-LINE FILTERS

A large number of power filters are available on the market. They consist of series inductors and shunt capacitors. The capacitors terminate on the metal case of the filter. In three-phase systems the three-phase leads and the neutral are filtered. It is illegal to filter the equipment grounding conductor.

FACILITY POWER FILTERS

These filters are often associated with power switches, "on" indicators, and power entry hardware. This arrangement is compact and saves the designer a great deal of time.

Power-line filters are designed to be effective at frequencies from 100 kHz to perhaps several hundred megahertz. Filters that are effective at lower frequencies require more components and are larger and more expensive. The filters used for a screen room are often physically very large. This is the result of providing filtering action at low frequencies.

To measure the performance of a filter over its full frequency range requires a valid setup. All test equipment and the filter must be bonded to a ground plane. The driving signals must be well-shielded to avoid any direct coupling. Bench testing using clip leads and portable test equipment cannot be effective in making measurements above a few megahertz. The filter is only effective relative to the case of the filter. If connections to the case are too inductive, the filter cannot function.

Power-line filters should be installed by bonding the filter case to a bulkhead. The unfiltered power conductors should be kept on the outside of the equipment being protected. If the unfiltered leads enter the equipment to get to the filter, they will radiate into the equipment bypassing the filter. The filter can be mounted flush with the equipment surface and still be electrically outside of the equipment. The bond between the filter case and the bulkhead should provide a low-impedance connection to the bulkhead. A snap-in filter does not meet this bonding requirement. A few mounting screws may also be inadequate. A bond that is effective at 100 MHz is a prepared surface connection free of paint or anodization.

All leads entering an enclosure can bring in radiation. This includes signal leads as well as power leads. It is ineffective to filter some leads and not others. The power filter is effective in stopping interference that would otherwise enter through the transformer. If other parallel paths are not blocked, then the filter will serve no purpose.

The NEC does not allow a filter to be placed in the equipment grounding conductor. This means that this conductor must terminate on the outside of the enclosure. A valid filter installation is shown in Figure 9.3.

9.12 FACILITY POWER FILTERS

The transformers used in a computer power center can limit common-mode interference from entering secondary power circuits. The shielding in the transformer is not perfect and some interference can get through.

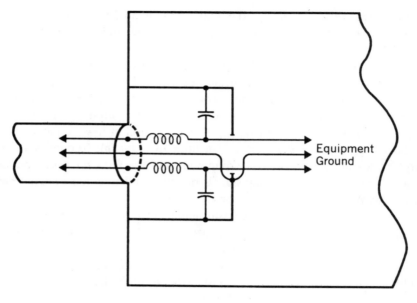

Power Line Filter — Correct
"Green Wire" Treatment

Figure 9.3 A valid filter installation.

To extend the isolation performance of the transformer, a power-line filter is often included in the package. The filter is mounted on the ground plane along with the transformer. The transformer shielding is effective at lower frequencies, and the passive filter is effective above about 100 kHz. If the power center is not mounted on the ground plane, any filtering that is provided is reduced in its effectivity.

9.13 TRANSIENT POWER LOADS

When a switch closes on a power circuit, energy flows to the load. This energy must come via fields over a transmission line from the generator. At the moment of switch closure there is no energy available and the voltage may drop to near zero. A step function of low voltage begins to propagate in both directions from the switch. In the power direction, energy is stored in various capacitances. Some of this energy moves toward the new load, and the voltage at the load begins to rise. As time progresses, a request for energy finally reaches the generator and the output is slowly increased. This phenomenon is complex because the transmission lines that are involved are certainly not simple. The request for energy can be observed on the power line at various distances from the load. The waveforms near a nonlinear load can be badly distorted. At a building entrance this distortion is reduced. At the pole transformer the distortion may not be observable. Energy is not created along the power line. It must come from the generator, and this takes time.

The power circuits in a facility are usually in conduit. If radiation is to take place, it will be on open wire or out of panels. Some radiation can occur whenever loads are switched on or off. Most of the noise that is coupled from the power conductors flows in capacitances that couple interference from the primary to secondary circuits. This interference results when energy demands are not met locally, and step voltages must follow the power conductors. If the requested energy is supplied locally, then the request for energy that propagates back to the generator is low frequency in nature. If the load is shut off, then the energy in process can be absorbed locally. This must last until the generator can be told to reduce its output. This is the purpose of a filter. It should supply energy and absorb energy locally so that high-frequency phenomena need not propagate back to the power source.

9.14 CONTROLLERS

A controller is a device that allows power to be taken from the power line over a fraction of the cycle. Controllers are used to regulate lighting, control motor speeds, or control motor torque. Controllers use switching devices, signal-controlled rectifiers (SCRs), or triacs. These switches are turned on at some firing angle and turned off at a zero-current crossing. Controllers can be single- or three-phase and can switch current levels greater than 400 A. If not handled correctly, switching transients can disrupt electronics in an entire facility. Even if the power is taken from a separately derived system, the radiation that can result can be a problem.

The filters for a controller should be an integral part of its design. A filter should supply energy to the load from a local capacitor. This voltage supplied to the load should rise in about 100 μsec to limit radiation. In a good controller design the high-frequency fields that are generated by the switch should be limited to the area of the controller.

9.15 TRANSIENT PROTECTION

If lightning enters a facility, it can couple a pulse to the power conductors. Because power conductors run in parallel the coupling is apt to be common mode in nature. Transient voltages can occur between power conductors and other grounded conductors in the facility. To limit the chance of damage, transient suppressors can be placed across the line and from power conductors to equipment housings.

The first line of defense in at the power entrance where the neutral (grounded conductor) is earthed. If a pulse passes this point, then the next line of defence can be metal oxide varistors (MOVs) placed between the power conductors and an associated panel housing. A varistor conducts above a specified voltage, and this limits the peak voltage on the line. These varistors can accommodate very high surge currents. They may have to be replaced after several hits because they can lose their effectivity.

Transient protection can be purchased as a component. The package consist of gaps, series inductors, gas discharge devices, diodes, and shunt varistors. The key to protection is to provide a low-impedance path to earth or to the reference ground plane for the excess voltage. Protection, if needed, should be provided in stages. It is unwise to limit the protection to just the end device. Gap protection on dc circuits can be a problem. If the gap conducts, there may be no mechanism available to extinguish the arc.

Transients that result from inductive "kicks" are often high frequency in nature because of arcing (see Section 8.11). These pulses have very steep wave fronts. This type of pulse must be slowed down so that the transient protection device can begin to respond. This is one reason why series inductors are necessary. Every inductance has a limiting resonant frequency. This means that if the pulse is fast enough, some of the energy can cross the inductance through the parasitic capacitance. This lacking can be covered by using a high-speed diode that can conduct within nanoseconds. Diodes cannot dissipate more than a few watt-seconds of energy before they are damaged. This is the reason why there needs to be a parallel path to dissipate the bulk of the pulse energy. It is obvious that transient protectors can be complex devices. If they are needed, they should be purchased from a reliable source. The effectivity is greatly dependent on the bonding of the device to a suitable ground. If they are not mounted correctly, they may be ineffective. Transient protection is usually provided as an integral part of a power control center.

9.16 UNGROUNDED POWER

Under special conditions the code does permit ungrounded power. On good example is in hospital operating rooms where the fear is of a gas explosion if there is an arc. This power must still be routed in grounded conduit. Receptacles supplying floating power must be identified with a delta and colored orange.

The navy uses ungrounded power on board ship. This reduces the risk of corrosion in salt water environments. During battle if there is a fault condition, the power will not be interrupted. If a fault is detected, it should be repaired as soon as practical. For electronic equipment a floating power source can be very noisy. An added transformer can be used to provide a grounded power source for these sensitive loads.

In some industrial environments where qualified personnel are present, floating power is permitted. If there is a fault in an electric crucible, power must be supplied until the load can be poured off. If this not done, the crucible may be lost. The power must still be routed in grounded conduit.

Low-voltage circuits must be grounded if supplied from a transformer connected to utility power. A fault in the transformer could make the secondary hot and result in a safety hazard. If two low-voltage circuits are cascaded, then the second circuit can be floating. The low voltages provided inside of hardware are not covered by the NEC.

9.17 THE NEC AND POWER CONNECTIONS

Engineers should be aware of certain code restrictions as they relate to power connections. Color coding of power conductors should be maintained. Neutral or grounded conductors should be white. Safety conductors should be bare or green. Conductors with green stripes should not be used except as equipment grounding conductors.

The code does not allow the equipment grounding conductor to be used for any other circuit function. It must be present to provide fault protection at all times. The code does not allow filters, inductors, or resistors to be placed in this line. Switching the equipment ground is not permitted.

The code does not permit the modification of listed hardware. Drilling holes or adding brackets is not permitted. The code seems to be somewhat arbitrary in some of its restrictions, but experience has made the rules necessary. The code rarely provides a rationale for a rule. Sometimes rules can be interpreted in more than one way. Cities often have their own code that stems from the NEC. City codes are not always up to date, and this can cause problems for some manufacturers. The codes supplied by the city are interpreted as law. Unfortunately, inspectors often interpret the code differently. It is necessary to follow these rules and their interpretation to avoid lawsuits. Insurance companies will look for someone to blame, and it should not be the engineer.

The power we use every day can burn down a building. The protection that is supplied by the power engineer must consider every contingency. Modifications to the power system must not interfere with the existing fault protection systems. This is the reason why changes must be made by qualified personnel and approved by city inspectors. Most interference problems can be handled without modifications to the power system. This should be the rule for the electronics engineer.

9.18 BENDING MAGNETIC FIELDS

The magnetic fields from power sources can be a source of trouble. Shielding against these fields (induction fields are low-impedance near fields) is difficult. It is possible to provide attenuation by using layers of steel and copper to surround the volume of interest. If the magnetic material has high permeability at low flux levels, the shielding effectivity is improved. In screen room designs the use of thick steel walls is required to provide any reasonable degree of shielding effectivity at power frequencies (harmonics).

Mu-metal is an exotic magnetic alloy that is often used in critical applications. It has excellent permeability at low flux levels. Special mu-metal covers are often placed around computer monitors to limit interference from nearby magnetic fields. Mu-metal is annealed in an inert atmosphere in a magnetic field to align the magnetic domains. The magnetic properties are lost if the part is drilled, punched, bent, or even dropped. For this reason the part must be formed before it is annealed. The largest size part is a function of the ovens that are available. Mu-metal parts are expensive because of the materials used, the special annealing, and the special handling.

Mu-metal is ineffective for flux levels above a few kilogauss. A mu-metal monitor cover works because the nearby magnetic flux follows the provided magnetic path around the monitor rather than pass through the monitor. Stated another way, the magnetic field reconfigures itself to store

BENDING MAGNETIC FIELDS

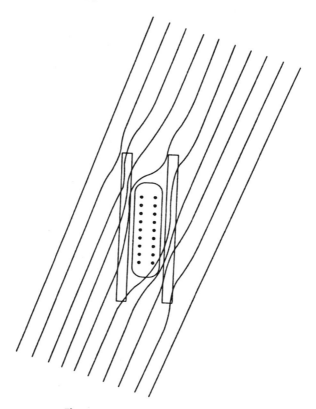

Figure 9.4 Bending the magnetic field.

a smaller amount of field energy. The field is not reflected, but it is simply "bent" out of the way.

This same principle can be used to keep field lines from crossing a critical signal path. This is shown in Figure 9.4, where the field might couple into a signal loop areas formed by the pins in a connector. Two small bars of steel warp the field enough to reduce the coupling in this critical area.

Cables can be wrapped with a flexible tape containing a layer of powdered magnetic material. These tapes are available commercially. Here again the idea is to deflect the magnetic field around the cable rather than have it penetrate the cable and introduce differential interference. This magnetic treatment is often provided in addition to standard metallic shielding. A magnetic shield need not be grounded to be effective.

10

HIGH-FREQUENCY DESIGN

10.1 INTRODUCTION

The design of high-frequency circuits began with the first radios and the first radio transmitters. The early designers found out about parasitics, high-frequency feedback, the tuning of transmitting antennas, the characteristics of transmission lines, the earth as a conductor, and the nature of radiation. Fortunately, there was little electrical interference and a radio receiver was a simple device. If it worked at all, it was sold.

Today, many of the practices used by the pioneers must be relearned. Our digital world is a high-frequency world. The physics doesn't change—just the components and the proliferation of hardware. Many a designer could learn a great deal by talking to a transmission engineer, but they are few and far between. In today's world of specialization the analog designer is almost extinct and the digital designers are mostly software oriented. There are a few amateur radio people around that can build their own equipment, but in the main there is a large gap between the practice of building high-frequency equipment and the new designer who sets out to manufacture a product.

At high frequencies the problem is almost all geometry rather than circuitry, although circuits are still an outline of what needs to be built. The simplistic ideas of power supplies and connecting components together falls apart as the frequencies rise. An AM radio is easy to build compared to an FM radio, and this is easy compared to building a tuner for a TV set. Without experience these designs cannot be built by simply hooking up components.

A large number of products revolve around printed circuit technology. It is very convenient to interconnect a group of integrated circuit (IC) components with printed wiring. It lends itself to mass production, and the hardware keeps getting smaller. The design techniques work, but the reason for all of the success may not be appreciated. Designers often take something that works and extend the art without asking questions. In the last decade, clock frequencies have gone from 5 MHz to 300 MHz and the designs look similar. If highway designers were to consider a factor of 2 increase in speed, the highways would have to be considerably different to be safe. A factor of two in electronics is tossed off as no problem. A factor of 10 increase in speed for autos is jet aircraft speed. What happens on a PC board when clock speeds increase from 100 MHz and 1 GHz?

10.2 THE PC BOARD

Two-sided boards allow traces to crisscross and interconnect components. Power supply leads can be interwoven with signal leads. The sheer number of interconnections limits the practical component density that can be accommodated. This technique works well for clock frequencies that are below 10 MHz and where the boards are relatively small, say 6 in. × 10 in. This type of board might mix analog and digital processes. It is rare that a pure analog design would require a multilevel board.

Digital speed and the interconnection problem have forced the development of multilevel boards. Today it is not uncommon to see boards with more than eight layers. The use of computer programs make it possible to provide very accurate film negatives. This film control has made it practical to maintain hole and trace registration patterns to within a 10-thousandth of an inch. This still takes careful attention to detail, including temperature and humidity control. When eight layers (four boards) are laminated together, the total thickness is held to 60-thousandths of an inch. The multilevel board allows for the use of power and ground planes. The outer planes can be ground providing shielding. Without multilevel construction a ground plane must be constructed by a grid of conductors that forms a pseudo third layer. This solution is labor intensive and is not as good as a thin sheet of copper on the board. Designers have tried the technique of routing ground on the perimeter of the board so that it does not interfere

PC BOARD TRANSMISSION LINES

with signal interconnections. The loop areas that result limit the speed and performance of these circuits.

10.3 PC GROUND PLANES

The ground plane on a PC board is used differently than a ground plane for a room. The PC board plane is used as a signal return conductor as well as a reference conductor. Every signal lead forms a transmission line with the ground plane. These transmission lines are short, and terminations are a poor match for their characteristic impedance. This means there are multiple reflections on every logic line. Because of the very small loop areas that are involved, the individual transmission lines do not radiate significant energy.

The fields that surround a logic line are most intense directly under the trace. The amount of field energy that is available to couple to an adjacent trace is small. This means that logic lines rarely cross-couple enough signal to upset any logic. For long lines this coupling issue may require further consideration.

10.4 PC BOARD TRANSMISSION LINES

It is worth noting the nature of the fields that transport logic signals from one component to another. When a logic switch closes, the energy must come from a local energy source. If this energy source is not provided, then the request for energy must propagate back to the power supply. If the forward and reverse paths are both 50-Ω lines, the resulting forward voltage is simply one half. A high-impedance termination allows a refection to raise this voltage. The standard way to provide for local energy is through the use of decoupling capacitors. These capacitors are located near the ICs and are connected between the power conductors and the ground plane. The number of decoupling capacitors depends on the speed of operation and component density. Typical capacitors might have a capacitance of 1000 pF. One capacitor might support the energy requirements for 4 ICs. Larger capacitors are not recommended because their natural frequencies are too low. Because of size the lead lengths alone add series inductance. This inductance adds time to the supply of energy. If more energy is required, then more small capacitors should be provided. Many solutions to the decoupling capacitor problem exist. Single-in-line package (SIP) arrays are available that solder in next to an IC. Capacitor groups can be mounted under the IC itself. Because of their small size and ease of mounting in mass production, surface-mounted capacitors are preferred.

If 30 logic switches operate at the same time and there is insufficient local energy, the power voltage sags and a wave propagates back to the power supply. A reduced voltage is now applied to every IC on the board. This time of propagation limits the clock rates that can be used. Transient phenomena must settle before the next clock time or there will be logic errors. It is always a good idea to check for signals on the power supply leads to make sure that high-frequency energy requests are not taking this path.

10.5 PC BOARDS AND RADIATION

The request for energy after a logic closure causes current to flow in two loop areas. These loops involve the capacitor leads and the IC leads. These loops can radiate energy (see Section 3.12). Obviously, surface-mounted components greatly reduce these loop areas. The worst-case radiation can be roughly calculated by considering the frequency of interest as $1/\pi \tau_r$. The rise time and voltage depends on the class of logic that is involved. The radiation is directly proportional to the loop areas, the number of gates firing at any one time, and the logic voltage. Use these parameters along with data from Figure 3.7 to calculate the worst-case radiation. This radiation is not attenuated by the ground plane as the components are mounted above its surface. A board might have 50 ICs and five gates firing per clock time. The coherent radiation from 250 sources can be a problem.

There is some radiation from external board traces. This results because the fields are not totally confined to the board. Multiple reflections occur because the lines are not properly terminated. The round-trip time for these reflections is not related to the clock frequency or its harmonics. The resulting radiation is broad band and not structured around harmonics of the clock. On a board with a thousand interconnections the radiation from any one line is small but when the effect is multiplied by a thousand, the radiation can be of concern.

10.6 PC BOARDS DECOUPLING CAPACITORS

Decoupling capacitors do not release their energy in zero time. Some of the charges can be trapped in the dielectric and the release time can be

PC BOARDS—GROUND PLANES AND POWER PLANES

extended. In high-speed circuits the type of dielectric that is used for decoupling capacitors is important. Just having a capacitor for decoupling may not suffice.

A very thin dielectric between the ground and power planes has been used to form a continuous energy source for high-speed logic. This buried capacitance is about 500 pF per square inch. Depending on circuit density and speed, two ground-power planes may be necessary. The fabrication process is licensed to many PC board manufacturers. This method of decoupling eliminates the need for discreet capacitors and associated "through" or "via" holes. This in turn allows for a smaller board with more traces per inch. For specific large scale integrated circuits the manufacturer may require local energy storage greater than that provided by the buried capacitance. Note that the ground and power planes can be viewed as a transmission line. The characteristic impedance can be considered below 1 Ω. The wave that propagates radially and reflects is complicated by loading and board geometry. (See Section 3.2).

10.7 PC BOARDS—GROUND PLANES AND POWER PLANES

The power plane and ground plane can be used to shield signals on separate layers. The fields for transmission are located between the traces and the ground plane. The fields on one side of a ground plane do not cross over to the other side. This means that there will be no cross-coupling of signals located on the opposite sides of a ground or power plane.

A power plane is a ground plane except that it has a dc voltage. It is a perfectly good shield and could be used as an outer layer. It is preferable to keep this plane as an inside layer to avoid the chance of a power short circuit if something accidentally contacts the plane.

Multilayer boards are expensive, particularly when a few boards are needed. Designers try hard to get by with a two-layer board. When a board fails to perform, adding ground plane to open spaces is often suggested. An examination of this approach indicates the fallacy. The ground plane is broken up by leads that interconnect components. Some of the signal loop areas that are formed can be many square centimeters. These areas are a far cry from the area formed by a trace over a ground plane. These are low-speed circuits and are susceptible to interference from external radiation.

Ground planes allow leads to take any path, and a ground return path is always provided nearby. If a ground plane is not available, it may be possible to route a ground trace (or power trace) next to every signal path.

This is cumbersome and requires many more traces. An external ground plane near the traces often can help. The fields from the traces reflect from the ground plane. This image reflection tends to cancel the initial field. The ground plane need not be grounded to form the image field. Fields that couple to exiting cables can be modified where the ground plane connects to the circuit. Obviously the nearby ground plane can also function as a faraday shield.

10.8 PC BOARDS—RIBBON CABLE

A ribbon cable is a flat arrangement of conductors like the keys on a piano. The individual insulated conductors are held together by a plastic bond. Ribbon cables are compatible with a wide range of connectors that can connect PC boards to other points in the equipment. This hardware is mass produced and is used in most product designs. Fortunately, in most applications the path length is short. The very tight field control available on the board is no longer available with the cable. The loop areas involved in signal transport are the issue. These loop areas can create problems in both interference and radiation. Fortunately, logic is very forgiving and in many applications there are few problems. The only design problem may be one of meeting published radiation specifications. It is important to understand best practice so that if the application is difficult, the problems can be avoided.

It would be desirable to extend the ground plane along the ribbon cable and connect it to the ground plane at the termination. Ribbon cable with a ground plane can be purchased, but the connectors to accommodate the shield are another story. The ground plane must be peeled back to allow a displacement connector to be mounted. The issue is then how to connect the ground plane to the PC boards. If it is connected with one conductor, the ground plane is violated. A ground plane is a geometric concept, not a circuit conductor.

A good practice in ribbon cable connections is to provide a ground return path (G) next to each signal lead (S). At high frequencies the power conductor (P) can serve as a return path. Every third conductor should be a return path. A typical wiring configuration might be SGSSGSSPS. The field for each signal uses the nearest ground. Each ground conductor must make a separate connection to the ground plane at both terminations. These grounding configurations limit radiation and susceptibility.

HIGH-FREQUENCY TRANSPORT AND OPEN CABLE

When ribbon cables cross an open area, common-mode coupling can occur. Common-mode signals are coupled by fields propagating in the space between the ribbon cable and any nearby ground plane (see Section 4.11). It makes sense to route cables on the ground plane to avoid this type of coupling. If designs provide mounting clips, they will be used and the chances of interference will be reduced. Coiling excess ribbon cable is not recommended. The chances of coupling to interference is obviously increased.

The spacing between leads on a ribbon cable is finite. In a strong field, this spacing permits differential coupling. To limit this coupling, pairs of conductors in the ribbon cable can be twisted. The signal coupled to each half-loop cancels the signal in the adjacent half-loop. Ribbon cable with twisted pairs can be specially ordered.

Twisting pairs of conductors can also be used to reduce the coupling between adjacent conductor pairs in the ribbon cable. It is preferable to twist every other pair. Twisting adjacent pairs in opposite directions may not be effective. Note that coupling depends on adjacent loop areas, not on how the areas are formed. Twisting with a different pitch can work, but this is a very special cable.

Ribbon cable is available with two ground planes, with coaxial shields around each signal conductor, and with these shields insulated from each other. These more elaborate ribbon cables are expensive and are usually not required. The problems of termination must be solved to make these cables useful. Connectors that provide a few coaxial connections are available. The ribbon conductors must be individually soldered.

10.9 HIGH-FREQUENCY TRANSPORT AND OPEN CABLE

Telephone wire pairs and open twisted pairs can be used to transport high-frequency signals. Some signal preemphasis and/or postemphasis may be required to accommodate line losses. Open bell wire has been used successfully to carry quality color video signals up to one mile. With repeaters the signal can be carried 10 miles or more. The insulation around the conductors plays a role in the line losses, and not all cable is practical. The line can have no branch circuits or intermediate terminations. The effective bandwidth for quality video is 10 MHz.

Computer memories in large computer installations have been accessed using twisted pair conductors. Clock rates of 100 MHz have been used, and the signals carried a distance of 60 ft. Some care must be taken to limit reflections at the receive end of the cable. The savings over using coaxial cable is obvious.

10.10 HIGH-FREQUENCY TRANSPORT OVER COAX

Many different cable types are available to transport signal and power at high frequencies. The losses in the cable as a function of frequency are provided by the cable manufacturer. For short runs the cable type is often not critical. Cables intended for high-frequency transport are often used in analog applications. Aluminum-foil-type cables should be avoided in high-frequency applications (see Section 6.10).

For flexibility, most cables make use of tinned copper braid for the return conductor. The braid density varies with cable type. A thick braid or a double braid provides additional attenuation against external fields.

Two factors limit the performance of coaxial cables at high frequencies. First, the dielectric used between the center conductor and the shield absorbs energy in a nonlinear way. This effect is called *distributed dielectric absorption*. Second, the surface roughness inside the shield causes a distributed reflection to occur. The cable TV companies have a unique cable problem. They must run miles of cable with low losses and very little reflection. The cable that is used in this application has a shield that is extruded aluminum, and it has an almost mirror inside surface. The center conductor is stiff, and it is centered by a spiral of nylon. The dielectric is basically air, and problems with dielectric absorption are minimal.

10.11 CABLE TRANSFER IMPEDANCE

When an electromagnetic field uses a cable to transport energy on the outside of the cable, current flows on the external shield. Ideally, this current flows on the outside surface of the shield and does not enter inside of the cable. Braid by its very nature can carry some of the surface current to the inside of the cable. Current on the inside surface implies a field that can propagate in both directions inside the cable. If the cable is properly terminated, a voltage V will appear at both terminations. The ratio

$$2V/I = Z \tag{1}$$

CABLE SHIELD TERMINATIONS

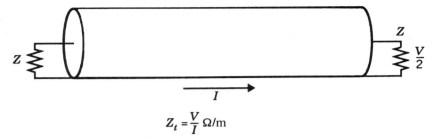

Figure 10.1 Transfer impedance.

is called the *transfer impedance*. The factor of two is used because half of the voltage appears at each termination. The measurement is defined for a cable length of 1 m (see Figure 10.1).

The transfer impedance for typical cables is shown in Figure 10.2. At low frequencies the current uses the entire shield, and the transfer impedance is the dc resistance of the shield. As the frequency rises, skin effect begins to enter the picture. For solid wall tubing the transfer impedance is very low at high frequencies because current cannot penetrate the shield. For braided cable the transfer impedance is greatest at about 100 MHz. In some cases a shielded cable is just as bad as having no shield. The cable needs to be properly selected to do a good job of interference rejection in this frequency range. When double braid is involved, the two braids are tightly coupled together by capacitance. The transfer impedance is slightly improved with two shields, but not to the degree expected.

Solid wall shielding is inflexible. Corrugated shields are available on some small cables. The corrugation allows for some needed flexibility. Flex used in power wiring can be very effective at high frequencies if the flex is of a type to provide a fluid seal. It should not be used for long runs because the inner conductors do not have a fixed geometry.

10.12 CABLE SHIELD TERMINATIONS

The subject of apertures was discussed in Section 4.6. Field energy can enter an aperture based on the dimension of the aperture and the half wavelength. When a conductor threads through an aperture the story is one of coaxial transmission which can occur at all frequencies. A cable and

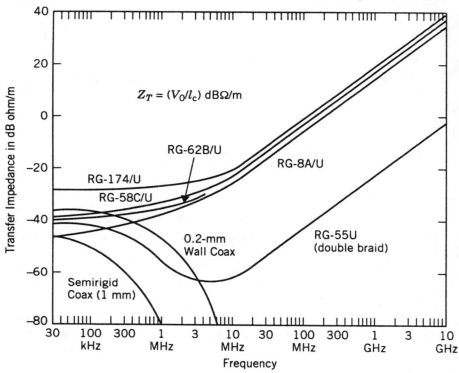

Figure 10.2 Transfer impedance of various cables. Courtesy of emf-emi Control® Inc.

connector, when mounted on a bulkhead, provides a possible path for interference if the aperture is not treated correctly.

Consider a cable shield that bonds to the bulkhead. Interference currents following in the cable will flow in the bond. If the bond is made by connecting the braid to the bulkhead at a single point, the fields around the connection can couple to the leads inside of the cable. The interference can then propagate on these leads into the hardware. This interference coupling can be far worse than the coupling allowed by the transfer impedance. Figure 10.3 shows a cable entry with various shield terminations. The worst possible case is when the shield is brought through the connector and terminated on the inside of the bulkhead. When the entire braid is used for a connection, the problem is still not solved. The correct solution is to provide a backshell for the connector. The shield is now bonded to the bulkhead for 360°. The issue is one of geometry and not circuit.

THE CHATTERING RELAY TEST

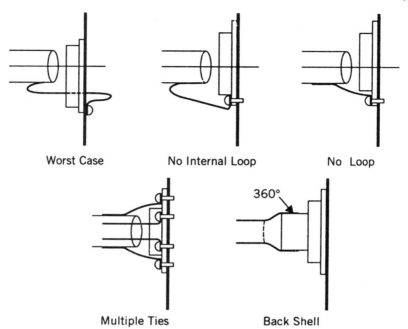

Figure 10.3 Cable shield terminations.

The problem just discussed is apt to occur on multiconductor terminations. The use of BNC or TNC coaxial connectors provides a good shield bond. If the connector is mounted on a painted or anodized surface, the bond may not be adequate.

10.13 THE CHATTERING RELAY TEST

Equipment is often mounted in harsh electrical environments. The power sources may be very noisy or there may be solenoids and relays in the vicinity. The military often requires equipment to function without error in this type of environment. The weak point in many installations is the cabling that interconnects hardware. This cabling is often routed next to conductors carrying very noisy signals. The hardware inside of a metal housing is usually not an issue. The test that has been devised is severe. A relay coil in series with its own contacts forms a buzzer. The series circuit is wrapped around the cable to provide transformer action. The number of turns per foot is specified. The interference that is coupled is very intense and covers a broad spectrum. Cables that do not terminate in backshell connectors will probably allow equipment to fail this test.

10.14 THE BOND

Circuits perform based on the interrelationship between components. The circuits are not always the ones that are represented on paper. Performance is modified by the presence of parasitics. Stray electric fields are described as a parasitic capacitances. The generation of magnetic fields are described as stray inductances. The attempt is to quantify the phenomena in terms of the circuit elements so that an analysis is possible.

The concepts of circuit begins to break down completely when skin effect is involved. At high frequencies, currents do not penetrate into the conductors and the results are not easily described in terms of circuits. The bond between two conductors could be described in microhms as long as the entire contact area is used. When skin effect is involved, the current may not use the entire bond and the ideas of impedance are strained. When current is forced to flow to the center of a conductor or through a conductor to make a connection, the impedance is raised.

The only way the impedance of a conductive path can be explained is to find the current density at all points in that path. Two high-frequency examples can explain this point. On one circuit board a high-frequency transistor was soldered to the board. In manufacture it was decided to rivet the component to avoid the possibility of overheating. The riveted part did not perform correctly because current did not flow uniformly to the component. In a second example the surface of a tinned surface was very irregular. When the surface was smooth the desired performance was achieved.

The part of a conductor that carries no current is redundant. To lower the impedance of a conductor at high frequencies the surface area must be increased, not the cross-sectional area. Current flow cannot be restricted to a narrow passage or the inductance will be increased. This all runs contrary to the strong feelings we get about circuits and circuit connections. A bond at 60 Hz is much different than a bond at 60 MHz. This means that surface currents must be considered if a filter is to function at high frequencies. The currents flow on the inside surfaces of the filter case, not the outside surfaces. A bond made to the inside surface is more effective than a bond made to an outside surface. If current must flow around the edges of a case to find a path, the impedance is increased.

10.15 THE TRANSPORT OF HIGH-FREQUENCY POWER

Cable with two conductors (twinax) can be used to transfer power at high frequencies. The signal might be balanced at the source with a suitable transformer. The sheath is grounded and connected to the midpoint of the signal. This type of cable avoids some of the losses (radiation) experienced in coaxial transmission. The shield is not a power conductor, although there is shield current. At the load end of the cable a second transformer, called a *balun,* can be used to convert the signal from balanced to unbalanced. Note that the word *balun* is coined from the two words balanced and unbalanced. Twinax connectors are often used by analog engineers to connect analog cables having a twisted pair of signal conductors. Unfortunately, this connector often has no keying provisions.

10.16 MIXING ANALOG AND DIGITAL SIGNALS

The approach taken by many engineers is to use separate ground planes for analog and digital circuitry. The connection between the two ground planes is made at one point. Unfortunately, this single connection is inductive and all signal currents must funnel through this narrow path. This adds a common-mode potential to the signals that cross at this bridge.

It makes more sense to use one ground plane but place the leads and components so that the resultant digital and analog fields do not share the same space. This planning includes the connector pin assignments. Even with one ground plane, there can still be field concentrations and resulting potential differences. These potential differences can be rejected by using suitable forward referencing amplifiers (see Section 7.8).

10.17 THE SNIFFER

Many simple tools can be built to measure fields. The tools are not calibrated, but they can be consistent. A half dipole can be mounted on a metal plate and used as a receiving antenna. A coil of wire can form a loop antenna. A coaxial cable can carry these sensors to an oscilloscope for observation.

A sniffer is a piece of coax terminated in a small loop where the center conductor bonds to the shield. This geometry is shown in Figure 10.4. The sniffer senses the *H* field in the single turn. The coaxial shield reflects the

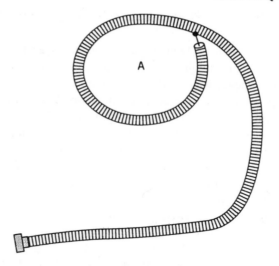

Figure 10.4 A sniffer.

E-field component. The H field can be calculated by using Faraday's law (see Section 2.5). The sniffer must be oriented such that the coupling is maximum.

The sniffer can be used to scan hardware for radiation leaks. It can be placed near apertures, connectors, seams, transformers, or PC boards to locate sources of radiation. The treatment of each leak involves better bonding, gasketing, wire mesh, or reduction of loop areas. These topics are covered in detail in Chapter 4.

10.18 SCREEN ROOMS

A screen room is a large metal box with power and air conditioning. The word *screen* was used in the early days of EMI testing when a wire mesh was applied to the walls, floor, and ceiling of a room. The wire mesh treatment does not meet today's needs for field attenuation. Shielding to be effective must consider power conductors, access doors, lighting, air conditioning, and a broad spectrum of interference. This has led to the use of solid wall enclosures.

A screen room can be used two ways. The equipment under test (EUT) can be operated in the screen room in a field-free environment. If there is

GROUNDING THE SCREEN ROOM

radiation, then the fields from the EUT can be measured. The fields in an urban environment are significant; and without a screen room, valid measurements are impossible. The second application involves radiating secure information. Equipment to decode and interpret secure information is placed in a screen room. It is necessary to make sure that field energy in any form cannot leave the room and be sensed. The concern for security involves the power source and the grounding scheme where there is fear that sensitive information might be accessed. This fear has introduced some strange practices that defy explanation.

10.19 SCREEN ROOM POWER FILTERS

Power-line filters are usually mounted on the outside wall of a screen room. All conductors except the equipment grounding conductor are filtered. The filtering is usually accomplished in stages to accommodate both low- and high-frequency requirements. The filter must be designed to contain the flow of surface currents on both sides of the filter. If internal surface currents can flow to the outside surfaces via the filter, then the integrity of the screen room is breached. Skin effect limits the flow of current to the surface at high frequencies but not at lower frequencies. This fact must be considered in the mounting of the filter and in its design.

Equipment grounding conductors cannot be filtered and they must terminate at the walls of the screen room. The filters must be built to provide filtering action for signals entering on the power leads and for signals that might be transported out on the power leads.

10.20 GROUNDING THE SCREEN ROOM

No additional grounding of a screen room is required. The equipment grounding conductor provides for safety and fault protection. If additional grounding is provided, surface currents can flow between the connections. At low frequencies some of this current can flow on the inside wall, and this limits the integrity of the room. If a ground is added, it should be at the power entrance to limit the opportunity for surface current.

The room is additionally grounded through capacitances on the floor. This capacitance can be as high as 0.01 μF, and at 1 MHz this is a reactance well below 1 Ω. For this reason, some screen rooms are supplied with a double conductive floor.

A ventilation duct would normally be bonded to the screen room. It is preferred to insulate the duct to at least 6 ft from the screen room. This eliminates another source of surface current. The 6-ft separation is suggested to avoid a possible lightning strike jumping to the screen room.

If fiber-optic cables enter the screen room, any steel support cable needs to be cut to avoid contaminating the room. Again the steel should be cut 6 ft from the room to keep lightning from jumping to the room from the cable. The fiber-optic cable must still be supported mechanically to avoid damage. Fiber optics can enter the screen room through a small-diameter tubing. The tubing acts as a waveguide beyond cutoff (see Section 4.9).

10.21 THE SCREEN ROOM DESIGN

The inside walls of the screen room need to be filleted. If surface current meets a sharp bend, the current will enter the wall. This means that some current can flow on the outside surface. To limit current from entering the wall the corners must be treated with welded fillets.

Doors are a weak spot in the design of a screen room. Finger stock can be used to seal the door aperture. This stock should be located inside a U channel to keep the fingers from catching on clothing. A gasket that is under pressure when the door is closed can also be made to work. The apertures must be kept small and the aperture depth must be maintained if the shielding is to be effective. The seal must look like a series of small apertures with long path lengths.

Signal cables must all be filtered at the screen room wall. The preferred entry point is near the power entry to avoid surface currents. Fluorescent lights are too noisy to use in a screen-room environment. Telephone lines must be filtered at the bulkhead. Lightning protection on telephone lines is usually handled at a distribution panel.

10.22 USING A SCREEN ROOM

Equipment mounted to the walls of the screen room will add to the surface currents that can flow. It is preferred that equipment be mounted in racks that are spaced away from the walls. If equipment generates low-impedance fields, it should be mounted as far from the wall as

USING A SCREEN ROOM

possible. Remember that low-impedance fields can penetrate the walls of a shielded room.

Power conduit should not be mounted on any outside screen room wall. The fields from the power conductors can penetrate the wall or cause surface currents. Power should be brought in at right angles to the wall. A power duct inside the room should be spaced away from the wall.

11

PULSES AND STEP FUNCTIONS

11.1 INTRODUCTION

Many types of interference are pulse-like in character. These pulses, sometimes called "glitches," can "ride" the power conductors into equipment. Some pulses are repetitive and related to power or its harmonics. Some pulses are generated at each cycle of any periodic process such as clock generation. Some pulses occur once, such as lightning or electrostatic discharge (ESD). The arc from a relay or solenoid opening can be considered a single pulse.

A fault condition can generate a single pulse. An example might be conductive fingers looking for a tear in a passing plastic film. When a tear is found, a pulse of current flows. The subsequent interruption of the current causes arcing, and this is a wide-spectrum phenomenon. The field generated by the arcing may not be easily contained. The field generated by the current can be contained if the loop area is that of a piece of coax.

A very intense pulse can be generated if an electron beam jumps to an improper return path. This condition might occur in a crucible heater. Any time power is switched on or off, a single transient pulse occurs. Arcing results if the load is inductive.

11.2 SPECTRUM OF A SQUARE WAVE

Repetitively, signals are said to have a Fourier spectrum. This spectrum is made up of harmonics of the fundamental signal. A continuous square wave can be shown to be made up of a set of sine waves. The lowest-frequency sine wave is related to the repetition rate. If the square wave is repeated every millisecond, then the fundamental sine wave is 1 kHz. The harmonics that make up the square wave are sine waves at 1, 3, 5, 7, 9, . . . kHz. The amplitude of each harmonic is inversely proportional to its harmonic number. If the square wave has a peak-to-peak amplitude of 2 volts, the signal can be represented by the expression

$$V = \frac{2}{\pi}\left[\sin \omega t + \frac{\sin 3\omega t}{3} + \ldots + \frac{\sin 5\omega t}{5} + \ldots \right] \quad (1)$$

Note that the peak amplitude of the fundamental is $2/\pi$. The square wave is shown in Figure 11.1.

Every practical square wave has a finite rise time. This can be explained as follows. Every voltage has an electric field. A change in voltage implies a change of energy stored in that field. A change in voltage in zero time

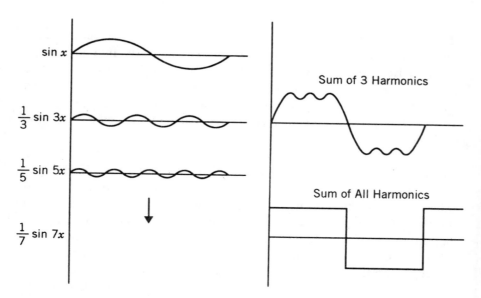

Figure 11.1 Square-wave harmonics.

SPECTRUM OF A SQUARE WAVE

FOURIER SERIES
Square wave with finite rise time

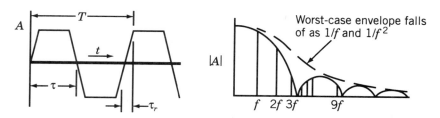

Figure 11.2 A typical square wave with finite rise and fall times.

implies that energy must be moved in zero time. This requires infinite power, which is not possible.

A Fourier analysis of a square wave with rise and fall times is more complex than an ideal square wave. The analysis shows that the lower harmonics that make up the square wave still have amplitudes that fall off proportional to harmonic number. Above a certain frequency the harmonics fall off proportional to the square of frequency. The mathematics is rather complex, and the solutions can be found in many textbooks. It is easy to see the impact of a finite rise time by plotting the envelope of harmonic amplitudes on a logarithmic scale. The square wave in Figure 11.2 is unsymmetrical with different rise and fall times. The harmonic envelope is shown in Figure 11.3. The envelope that is constructed is a worst case where no harmonic exceeds the envelope in amplitude. Note that the envelope has

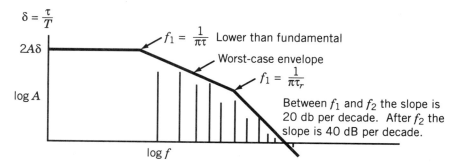

Figure 11.3 The envelope of sine-wave amplitudes that make up a general square wave.

two corners (three slopes). The first corner frequency is at $1/\pi\tau$. The second corner frequency is at $1/\pi\tau_r$ where

$$\text{Duty cycle} = \delta = \frac{\tau}{T} \qquad (2)$$

$$\tau = \text{time of one pulse} \qquad (3)$$

$$T = \text{period} \qquad (4)$$

In general, the fundamental (first harmonic) does not occur at the first corner frequency, and no harmonic occurs exactly at the second corner frequency. These two corners are just construction points.

11.3 SPECTRUM OF A SINGLE EVENT

It is interesting to consider what happens when the square wave in Figure 11.3 is modified so that τ is small. This waveform is shown in Figure 11.4. The harmonics and their worst-case envelope is shown in Figure 11.5. Note that the lower harmonic frequencies are below the first corner frequency and that the fundamental frequency is much lower. Another thing to notice is that the amplitudes of the harmonics are smaller. The envelope is a worst-case amplitude for each harmonic.

It is interesting to see what happens when the duty cycle is made zero and there is only one pulse. The result is that there is frequency content at all frequencies, and the concept of harmonics must be abandoned. The

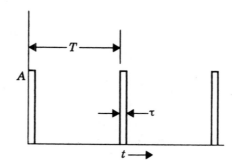

Figure 11.4 A series of short pulses.

SPECTRUM OF A SINGLE EVENT

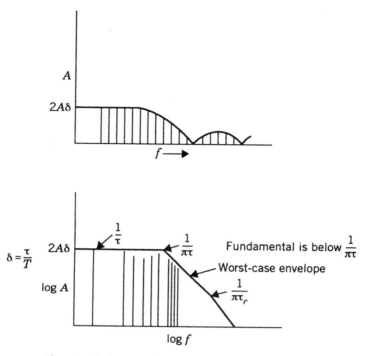

Figure 11.5 The envelope of harmonics when δ is small.

amplitude at any one frequency goes to zero, and a new approach must be taken in analyzing the spectrum.

The spectral envelope for a single pulse is shown in Figure 11.6. Note that the vertical scale has units of $2A\tau$, where τ is the time width of the pulse. The scale could also be marked $2A/f$ where f is the reciprocal of τ. The vertical scale indicates amplitude per unit bandwidth at that point. If τ is in microseconds, then the bandwidth units are in megahertz.

A single pulse can be idealized mathematically to have zero rise time and infinite amplitude. The area under the pulse can be set to unity. Such a pulse has a flat spectrum from dc to infinity. The important thing about this pulse is that there is content at all frequencies. This pulse is named after the mathematician Dirac. In the real world a pulse that has a first corner frequency well above the response frequency of the circuit can be used to excite every mode in the circuit. A good example is when impulse testing is used in testing mechanical systems. A sharp hit with a hammer can excite all of the modes of vibration.

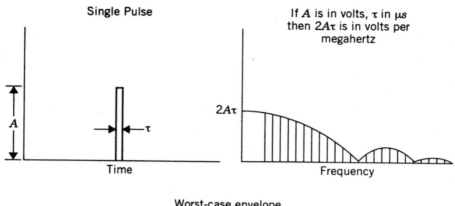
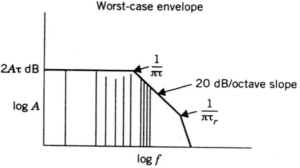

Figure 11.6 The spectral envelope for a single pulse.

The signal content in different bandwidths follows the rule of adding incoherent sine waves. When sine waves at different frequencies are summed, their root-mean-square amplitude value is the square root of the sum of the squares. This rule, when applied to bandwidth, means that the signal in twice the bandwidth is increased by the square root of two. If the spectrum amplitude is 40 dB microvolts per 1 MHz, it is 30 dB microvolts per 100 kHz and 20 dB microvolts per 10 kHz. These figures are again for one point on the envelope. If the envelope is flat in the region of interest, the signal in a given band can be easily calculated. If the band of interest is on the slope of the envelope, an integral is required. The envelope is an approximation, so an easy way to calculate the signal over a portion of the spectrum is to divide the spectrum into small sections and use the root-mean-square principle to add the signals together.

Spectrum analyzers allow the users to select filter bandwidth. The analyzer provides data by moving the filter across the band of interest. In some cases the filter bandwidth is automatically changed as a function of frequency. This change of bandwidth can appear as a step change in signal level.

11.4 A VALUABLE CALCULATING TOOL

When a pulse couples to a circuit, the resulting peak voltage is often of concern. Excess voltage might damage a circuit. In logic circuits the presence of a pulse during a clock transition can cause a logic error. Mathematical techniques are available to calculate the impact of a transient on a circuit. When the waveform is nonstandard, this method can be very cumbersome. Computer simulations can be used if enough information is available. If the harmonics are known, the response to each harmonic can be summed to determine the response. In a nonlinear system, this treatment may be invalid.

The exact circuit solution is often of little value. What is more important is a worst-case analysis. If this analysis shows that there is no problem, then the engineer can move on. A simple approach is to use a sine wave for analysis. The frequency chosen is the second corner frequency as used in the envelope generation. The sine-wave amplitude is that of the first corner frequency. The peak amplitude of the result is the worst-case voltage that will be coupled. The reason this method works is that in most coupling phenomena the coupling is proportional to frequency. From the first corner frequency to the second corner frequency the amplitude content falls off proportional to frequency. The end result is that the coupled signal is flat to the second corner frequency. This is not an accurate method, but it is simple. A 30% error is just 10 dB. In most designs a 20-dB margin is required to be on the safe side.

11.5 THE ESD (ELECTROSTATIC DISCHARGE)

ESD is a common occurrence. In a clothes dryer the arcing is easily visible. Anti-cling materials are often used in the drying cycle to avoid this nuisance. In the desert, where the humidity is low, an ESD pulse from a key to a door lock can be an annoyance. In a hospital an ESD pulse could ignite gases or upset the performance of monitoring hardware. In this environment an ESD pulse is dangerous. In large grain storage bins the motion of dust particles can result in a discharge that can cause an explosion.

In a computer facility an ESD can disrupt the computer. In the assembly of weapons an ESD could arm a missile. In the manufacture of integrated circuits an ESD could damage the product. Some of this damage might not be felt until months later in product application. This potential for disaster requires that facilities take great care to avoid having ESD problems.

11.6 ESD PROTECTION

The best protection against ESD is humidity control. If the humidity is above 40%, then few problems will result. Humidifiers added to a room can still leave pockets of dry air. The best solution is to have humidity control a part of the air-conditioning system for the facility.

In very sensitive facilities, conductive paint is applied to floors and operating surfaces. All movable carts are grounded through their wheels to the floor. The intent is to ensure that charges cannot build up on any rubbed surface. If a tray rubs on a surface that is grounded, charges are immediately dissipated.

A technique often recommended is to ionize the air in sensitive areas. This ionization allows charge to be neutralized on nearby surfaces. A very simple technique is to place an open flame near a critical operation. The gases that are emitted provide the required ionization. Ionization hardware is also available.

A common misconception relates to grounding hardware. When a human develops a charge, the arc is usually to a grounded structure. By grounding everything, an arc has more places to go. It is important to ground conductors, but it is also necessary to stop charge generation. Grounding alone cannot solve this problem. For example, if integrated circuits (ICs) are pulled from a plastic bag, a charge can be picked up on the ICs. Placing these ICs on a grounded table may destroy them if the table is grounded. As the ICs near the table, an arc can jump from the IC to the table.

11.7 ESD AND ITS CHARACTERISTICS

An ESD pulse is characterized at a frequency of 300 MHz and at a current level of 5 A. The near field is considered inductive. At 300 MHz the near-field/far-field interface is at a distance of about 16 cm. Manufacturers of computers are very much aware of the impact an ESD pulse can have on their hardware. For this reason, extensive testing is done to harden the designs so that hardware will not be damaged by an ESD.

Consider the ESD case where a human gathers charge on a rug. When the human approaches a door knob, most of the electric field energy just prior to the pulse is stored in a volume around the finger that extends 1 ft. The pulse rises in the time it takes for the field energy to travel 1 ft. Using the speed of light this time is about 1 nsec. This is the rise time used to calculate the second corner frequency in the spectral envelope. This explains the reason why 300 MHz characterizes the pulse.

An ESD pulse can enter a piece of equipment over any conductor or through any aperture. In most cases the discharge path is through the equipment back to earth. Conductors that might carry pulse current should not be routed near circuitry. The changing magnetic field can couple into any loop area and damage components. The current should be provided a wide path so that field intensity near the path is lowered. This can be a conductive painted surface or a wide strip of aluminum.

Keypads and keyboards must be treated so that the ESD path is not through signal conductors into the hardware. A conducting membrane behind the pad should be bonded to the bulkhead. Any opening in the bond is an aperture that can allow the entry of field energy.

Suppression diodes can be placed on specific conductors to limit the voltage. These diodes must be high speed if they are to provide protection. A small series resistor (10–100 Ω) together with device capacitance can also provide some protection. It is difficult to use a series inductor because the natural frequency can rarely be made high enough for the inductor to be effective. Ferrite beads (a one- or two-turn inductor) adds a very small series impedance and are usually ineffective.

11.8 ESD TESTING

ESD pulses can be generated by various testers that are on the market. They are known in the trade as *zappers* or *zapping guns*. These testers allow the user to adjust the voltage, the repetition rate, and the energy in each pulse. For a human being the storage capacitor is about 300 pF. There are often two modes to the tester. In one mode the arc is generated at the tip of the probe. In a second mode the arc is generated in the tester while the probe is held in contact with the hardware being tested.

ESD testing should proceed on an orderly basis. The test voltage should start low, say around 1000 V. The arc should be applied to suspected weak points in sequence. This should be done while the equipment is functioning, and any malfunction should be noted. If there is a problem, the testing should stop and some attempt should be made to correct the problem. If no problems are noted, then the voltage should be raised in 1000-V increments. The idea is not to damage the hardware during the test.

Most testers can supply pulses up to 12,000 V. The critical testing voltage is around 7000 V. Above this voltage, much of the pulse energy is lost in heat and sound and is not radiated. The important thing to note is that testing should not jump to 12,000 V and skip the mid-range. If there is going to be trouble, it will occur at around 7000 V.

The arc at the probe tip places a low-impedance field on the surface of the hardware. With a remote pulse a wave propagates down the test cable and there is some reflection at the probe contact while some of the current propagates into the hardware. This remote pulse method is a different test. It is wise to consider using both modes if they are available.

When hardware passes an ESD test, it is usually going to pass any other form of radiation testing. The reason lies in the relationship between susceptibility and radiation. If a circuit has large loop areas, it is apt to radiate, and also external fields can couple to the hardware. If these loop areas are under control, there is low radiation and the hardware is not prone to radiate. If there are apertures that allow radiation to enter, these same apertures will allow radiation to exit. In one direction it is susceptibility and in the other it is radiation.

INDEX

A/D converter, 84
ac amplifiers, 140
ac system:
 coupling capacitors, 141
Absorption:
 shielding, 66
Active bridge arm:
 mounting, 96
Active components:
 paralleling, 146
Active impedances, 87
Air gaps:
 in magnetic path, 29
Aircraft ground, 16, 154
Aliasing error, 139
Ampere's law, 27
Amplifier, charge, 137
Analog and digital mix, 103, 179
Analog cable, 104
Antenna gain, 159
Aperture design:
 screen room, 182
 closing, 67
Apertures, 66
Apertures, independant, 67
Arcing on contacts, 142
Attenuation, waveguides, 68

Audio system grounding, 118
Audio transformers, 69

B field magnetics, 26, 30
Balanced excitation supplies, 133
Balanced input on oscilloscopes, 91
Body skin resistance, 73
Bonding, 66, 174
 filters, 159
 ground plane, 153
Box-ideal, shielding, 75
Bridge configurations, 96
Buried capacitance, 171

CMR see common mode
CRT shielding, 68
Cable:
 analog, 104
 braid, 104
 cable TV, 174
 low noise, 106
 pyroelectric effects, 106
 turboelectric effect, 106
 twisting conductors, 105
Calibration, charge amplifiers, 137
Capacitance, definition, 9

Capacitance:
  between spherical shells, 15
  cylinders, 18
  mutual, 21
  parasitic, 21
  self, 21
  to shield, 94
Capacitance of earth, 14
Capacitive feedback, 76
Capacitivity of free space, 14
Capacitors, parallel plate, 19
Carrier systems, 91
Changing E field, resulting H field, 40
Characteristic impedance, transmission lines, 46, 49
Charge:
  forces, 4
  units, 4
Charge amplifiers, 103, 135
Charge distribution, 16, 23
Charges:
  on earth, 12
  spherical shells, 14
Chattering relay test, 177
Circuit in a box, 76
Clean ground, 2
Coax, 94, 98, 174
Coaxial cable, 99
Coaxial current path, 50
Coaxial transmission, 50
Common-impedance coupling, 95, 111
  amplifiers, 112
  earth, 120
  neutrals, 119
  output circuits, 112
Common mode, 73, 178
  frequency dependence, 93
Common-mode attenuator, 128
Common-mode conversion, 90
Common-mode coupling, 71, 107, 151
Common-mode and power, 157
Common-mode rejection, 127, 128
  post modulation, 128
Common-mode signals, 87, 122
  derived, 101
  neutral drop, 119
  types, 73
Common-mode vs average signal, 74
Compass needle, 25
Computer power, 157
Computer grounding, 155
Conductors, internal current, 61
Conduit, power, 161
Conservative field, 7
Controllers, noise, 161
Core saturation, 132
Counterpoise, 121, 154
Coupling:
  fields to circuits, 69
  loops, 70
Current on metal surfaces, 64
Current loops, 57
Current path:
  transmission line, 48
  high impededance circuits, 97
Current source, excitation supply, 115
Cylinders:
  capacitance, 118
  charge distribution, 17

D field, 9
  at boundries, 9
Decoupling capacitors, 170
  pc boards, 169
Dielectric absorption, 174
Dielectric constant, 5, 13
Difference signals, 91
Differential amplifiers, 91, 122
  balanced signals, 96
  common-mode, 92
Digital and analog mix, 179
Digital grounding, 118
Digitizing signals, 91
Dipole, wave impedance, 94
Displacement field, 9
Distributed parameters, transmission lines, 44
Distribution panels, signal, 95
Distribution transformers, 149
Drain wires, 106
Drain wires:
  cables, 105
Dual-mode power supplies, 135
Duty cycle, 188

INDEX

E and H fields together, 40
E field:
  charge distribution, 5
  concepts, 22
  definition, 8
  energy storage, 55
  lines of force, 10
  MKS system, 13
  points of difficulty, 12
ESD, 185
  characteristics, 192
  floors, 151
  protection, 192
  pulses, 25
  testing, 191, 193
Earth, 120
  conductor, 16
  connections, 2
  currents, 120
  field terminations, 12
  impedances, 120
  plane, 15
  resistance, 2
Earthing, safety, 2
Effective radiated power, 59
Electric field, *see* E field
Electric field, 5
  energy storage, 20
  vector field, 5
Electric flux, 5
Electrolysis, 122
Energy storage, inductance, 35
Energy transfer at dc, 54
Energy transport, 41
Envelope of spectrum, 187, 190
Equipment ground, 100, 150
  routing, 119
  switching, 163
Excitation supplies
  balanced, 133
  transformers, 116

Far field radiation, 57
Faraday shield, 172
Faraday transformer shield, 82
Faraday's law, 29
Fault current, 154
Fault paths, 152
Ferrite cores, 34, 142
Fiber optics, 91, 153, 182
Field coupling, 77
Field effect transistors, 41
Field energy, electric, 20
Field energy capture, 44
Field lines, 10
Field representations, 10, 72
Fields:
  components, 41
  conductors, 61
  in a room, 10
  neutral coupling, 72
  transformer coupling, 72
Filters:
  analog signals, 139
  bonding, 159
  digital data, 139
  equipment, 155
  input leads, 94
  interference, 94
  guard shield, 101
  inputs, 98
  line, 144
  location, 159
  power, 150, 158, 161
  screen rooms, 181
  spectrum analysis, 191
  state variable, 139
  testing, 159
Floating circuits, 97
Forces on chages, 4
Forward referencing amplifier, 122
Fourier analysis, 187
Fundamental instrumentation problem, 90

Gage excitation, 115
Gaps:
  energy storage, 143
  ferrite cores, 34
Gaskets, 68
Gauss's electric flux thoerem, 12
Green-wire treatment, filters, 160

Ground:
  driving, 93
  earth, 149
  ribbon cable, 172
  safety, 99
  traces, 168
  virtual, 135
Ground grid, PC board, 168
Ground loops, 86
Ground plane, 49, 103, 121, 123, 150, 169, 171
  charge distribution, 17, 21
  PC board, 169
  single-point grounds, 151
Ground plane charge distribution, 17
Ground traces, 168
Grounding:
  audio, 118
  building, 118
  equipment, 100
  excitation supplies, 114
  human body, 147
  neutral, 118
  one conductor, 78
  output leads, 86
  ribbon cables, 171
  screen rooms, 181
  service entrance, 117
  signal leads, 85
  signal reference conductor, 81
  single point, 117
  single-ended signal, 95
  star connection, 117
Grounding electrode system, 2
Grounding rule, 79
Grounding the circuit, 77
Grounding the neutral, 119
Ground loop, 95
Grounds:
  isolated, 154
  multiple signal, 88
Guard rings, 108
  building, 122
Guard shield, 89
  leakage path, 97
  potentials, 101
  ties, 90
  where connect, 90
Guarding:
  calibration, 138
  summing point, 109

H field, 27
H field, energy storage, 55
Harmonics, 187
Headroom, 98, 139
High-frequency design, 167
High input impedance circuits, 130
Humidity control of ESD, 192
Hysteresis in magnetics, 33

Image plane, 172
Impedance:
  bonds, 178
  common-mode rejection, 90
Induced charge, 15
Induced charge, current flow, 22
Inductance, 35, 152
  leakage, 37
  mutual, 35
  simple conductors, 42
  solenoid, 35
Induction:
  B field, 27
  Faraday's law, 29
Induction field, 76, 104
Inductive coupling, 107
Inductors, 142
Inductors, ferrite, 35
Input guard shield, 98
Instability, output stage, 94
Interference:
  control, 3, 23, 78
  coupling, 107
  input leads, 75
  magnetic, 25
  transformer capacitance, 82
Ionization of air, 192
Isolated ground, 154
Isolated output, 123
Isolation transformers, 85, 100, 156
  power, 156
  shields, 156
  transformers, 34

INDEX

Laminations, transformer, 14
Leakage flux, 37
Leakage paths, 97
Lightning, 25, 118, 121
   facilities, 153
   screen rooms, 153
Line filters, 144
Line unbalance, 90
Loop area interference coupling, 108, 145
Loop area field coupling, 108
Low frequency definition, 3

Mad cow problem, 148
Magnetic circuits, 29
Magnetic field:
   B field, 26
   bending, 164
   earth, 25
   energy storage, 36
   lines of force, 26
   transformers, 32
Magnetic flux, 29
   current, 26
Magnetic heads, interference, 31
Magnetic hysteresis, 33
Magnetic shielding, 165
Magnetic storms, 39
Magnetics, 25
Magnetics units, 37
Magnetizing current, 30
Medical problem, 147
Microphone cable, 94
MKS system, 12
Multilayer PC board, 168
Multiple grounds, 88
Mutual capacitance, 76
Mutual inductance, 35

NEC, 2, 120, 152, 154, 163
Near field radiation, 57
Neutral grounding, 118, 149
Neutral proplems, 156
Normal mode, 73

Off-line switchers, 100
Ohms-per-square, 62

Open cable, 173
Open wire transmission, 87
Oscilloscope grounding, 86
Overload on input stages, 97

PC boards:
   multilayer, 168
   radiation, 170
PC ground planes, 169
Parasitic capacitance, 21, 77
Parasitic currents in transformers, 144
Permeability:
   free space, 37
   magnetic, 27
   relative, 27
   transformer iron, 33
Physics, teaching, 3
Piezoelectric devices, 106
Point charges, 12
Potential difference, 7
Power:
   separately derived, 155
   connections, 163
   filters, 158
   planes, 171
   transport, 43, 179
Poynting's vector, 54
Prefilters, 98
Primary fault path, transformers, 83
Pulses and step functions, 185
Pyroelectric effects, 106

Quiet ground, 2

Radar signal, 60
Radiation:
   ESD, 56
   current loop, 57
   dipole, 56
   loops, 58
   PC board, 59
Reactive coupling:
   circuits, 107
   structure, 96
Rebars, 150
Recording heads, 31
Reference conductors, 73

Reference junctions, 98
Reflection:
 junctions, 98
 plane waves, 64
 shielding term, 66
 transmission line, 46
Remote sensing, 115
Ribbon cable, 71, 98, 171, 172
Rise time, 187

SIL PAD, 145
Sacrificial anode, 122
Safety, 152
Safety ground, 83, 99, 117
Screen room, 72, 180, 183
 field penetration, 183
 power filters, 181
Screening, apertures, 67
Self capacitance, 21
Separately derived power, 155
Service entrance grounding, 117, 155
Shielding:
 balanced sources, 96
 bond, 176
 box, 82
 connection, 95
 dipole, 57
 electrostatic, 23
 effectivity, 65
 excitation supplies, 116
 E fields, 76
 ferrites, 66
 guard, 89
 high frequency cables, 175
 instrumentation, 100
 local, 101
 low frequency signals, 93
 material, 82
 near induction, 76
 permeability, 66
 single-ended, 94
 transformer, 82
 unbalanced circuits, 98
Shields:
 aluminum, 105
 common-mode, 157
 driven, 103

 isolation transformer, 156
 multiple in transformer, 89
Signal conductors, twisting, 104
Signal errors, 197
Signal rectification, 94, 98
Signal reference conductor, 81
Signal sources:
 balanced, 86
 human body, 147
Single-ended instruments, 85
Single point grounds, 117, 151
Skin depth:
 definition, 65
 earth, 121, 154
Skin effect, 61, 64, 178
Skin resistance, 145
Sniffer, 179
Solenoid:
 flux pattern, 26
 inductance, 35
Spacecraft potentials, 40
Spectrum:
 single event, 188
 square waves, 186
Spectrum analysis, 190
Stability, parallel components, 147
Star connection grounding, 117
Strain-gage bridge, 88
Stringers, ground plane, 150
Summing-point guarding, 109
Sunspot activity, 121
Switchers, 100
Switching regulators, 142
Switching supply transformers, 144
Switching transistors, 145

Terminations, fractional, 53
Thermocouples, bonding, 98
Transducers, 86
Transfer impedance, 174
Transformer:
 audio, 39
 capacitance, 81
 common-mode, 91
 coupling, 128, 140
 equivalent circuit, 37

INDEX

excitation supply, 116
external flux, 37
gaps, 39
as inductor, 39
isolation, 85
laminations, 34
leakage inductance, 37
mounting, 38
shield ties, 83
Transient protection, 162
Transient response of filters, 140
Transients on power line, 160
Transistor switching, 145

Transmission lines:
characteristic impedance, 46
current path, 48
fields, 47
optimum power transfer, 53
PC board radiation, 47
reflections, 46, 53
resonance, 53
sinusoids, 50
speed of propagation, 45
terminations, 45, 53
vector field, 5, 26
video, 54
wave guides, 68
wavelength, 51